PITT LATIN AMERICAN SERIES **PLAS**

George W. Grayson

and Mexican
Foreign Policy

University of Pittsburgh Press

Published by the University of Pittsburgh Press, Pittsburgh, Pa. 15260
Copyright © 1988, University of Pittsburgh Press
All rights reserved
Feffer and Simons, Inc., London
Manufactured in the United States of America

Library of Congress Cataloging-in-Publication Data

Grayson, George W., 1938–
    Oil and Mexican foreign policy / George W. Grayson
        p.   cm.—(Pitts Latin American series)
    Bibliography: p. 191
    Includes index.
    ISBN 0–8229–3574–0
    1. Mexico—Foreign relations—1970– .   2. Petroleum products—
Prices—Mexico—History—20th century.   3. Petroleum history and
trade—Mexico—History—20th century.   Mexico—Foreign relations—
United States.   5. United States—Foreign relations—Mexico.
I. Title.   II. Series.
F1236.G76   1988
327.72—dc19   87–25177
                                                                                    CIP

To Dr. Carmen Brissette Grayson,
in appreciation of her twenty-five years
of understanding, sensitivity, patience,
and—above all—enriching love.

Contents

Tables

Figures

Acknowledgments

Like every author I owe a heavy debt of appreciation to the many persons who have helped in various ways in the preparation of this book. Above all I wish to thank David A. Dessler, my colleague at the College of William and Mary, for introducing me to the literature on national role conceptions and providing invaluable insights on how this tool could be applied to Mexico's recent involvement in the international system.

In addition, I am indebted to some fifty Mexican, U.S., Venezuelan, and Canadian officials whose lengthy, off-the-record interviews contributed greatly to my understanding of the relationship between oil and Mexican foreign policy. I only regret that the sensitivity of their positions prevents my individually naming these outstanding public servants.

For extremely helpful comments on portions of the manuscript, I wish to acknowledge the assistance of Judith Ewell of the College of William and Mary; John J. Bailey of Georgetown University; Roderic Camp of Central College; M. Delal Baer of the Mexico Project, Center for Strategic and International Studies; Bruce Michael Bagley of Johns Hopkins University's School of Advanced International Studies; Michael C. Meyer, professor of History and director of the Latin American Area Center at the University of Arizona; Ing. Walter Friedeberg M., formerly production manager for Petróleos Mexicanos and now a member of the engineering faculty of the National Autonomous University of Mexico; and Stephen Lande, formerly the chief negotiator with Mexico in the U.S. Office of Special Trade Representative and now vice-president of Manchester Associates, Ltd. Drawing on data supplied by the U.S. Department of State, Rodney G. Tomlinson of the United States Naval Academy was kind enough to generate percentages for the coincidence of U.S. votes in the United Nations' General Assembly with votes cast by other nations in the Western Hemisphere.

I also wish to thank Catherine Marshall, managing editor of the University of Pittsburgh Press, and Jane Flanders, who edited the manuscript, for the painstaking care and discrimination they devoted to improving my roughly hewn prose while enhancing the style, organization, and coherence of the study.

An earlier version of chapter 7 was published as "Nicaragua: Soviets Aid with Oil Supplies" in the July 1985 number of the *Petroleum Economist*. I appreciate the willingness of Bryan Cooper, the journal's editor and publisher, to allow the inclusion of this material in my book.

Years consumed in preparing a manuscript also create obligations of a financial nature. The College of William and Mary, through its Faculty Research Committee, awarded grants that permitted two summers and one semester of research, including three visits to Mexico.

My thanks go to Sharon Clarke, Barbara Wright, and Elizabeth Ann Mack, who with cheerfulness, patience, and forbearance typed the manuscript; to past and present William and Mary students Kathy Moriarty, Sharon Clarke, Celia Klimock, Rebecca Lawler, Cathleen Welsh, and John A. Hosterman, who diligently assisted with research, proofreading, and preparing tables; to Myron B. Hayslett, director of educational media in William and Mary's Swem Library, for designing the graphics used in the two figures; to Lois Schuermann, assistant librarian at the American Petroleum Institute, who helped me track down scores of articles pertaining to Mexican oil; to Georgette Magassy Dorn, specialist in Hispanic culture in the Hispanic Division of the Library of Congress, who made it possible for me to gain access to the speeches of Mexican presidents; and to Dr. Carmen Brissette Grayson, a historian's historian whose intelligence, curiosity, erudition, good humor, enthusiasm for life, and sensitivity have made her an ideal marital and intellectual companion for twenty-five years, who tendered searching insights, suggestions, and criticisms that sharpened my thinking about the subject matter. Our children, Gisèle and Keller, generously permitted their dad evenings at the office to complete the manuscript, and—when he was home—allowed him occasional access to the telephone to speak with other Mexicanists.

With all the assistance received, it is obvious that the shortcomings of the study are my own.

Oil and Mexican Foreign Policy

Introduction

To Mexicans, oil is more than a flammable bituminous liquid trapped in the earth's upper strata. Among other things, it is essential to the mystique of the Mexican revolution that erupted in 1910. Politicians continually invoke the memory of the revolution to justify the promulgation of preferred programs; the dominant political grouping is the Institutional Revolutionary party (PRI), whose very name is an oxymoron; and the designation "citizen president" adorns every official document of the chief executive—a testimony to his revolutionary credentials.

Such references epitomize the symbolism that has long since eclipsed substance in the revolution. Indeed, the poor who live in destitution at the base of Mexico's increasingly squat social pyramid have seen their share of national income shrink even amid the post–World War II economic advances and the oil boom of the late 1970s and early 1980s. Moreover, the purchasing power of factory workers has plummeted in recent years and land distribution to peasants has long since ended—with the exception of an impulsive, politically motivated property seizure by President Luis Echeverría Alvarez just before he left office in 1976. That Echeverría's successor, José López Portillo, nationalized Mexico's fifty-four private banks three months before his term ended in 1982 appeared as another attempt to emulate the revolutionary actions of President Lázaro Cárdenas and not as evidence of a new revolutionary spirit suffusing his nation's political system.

In fact, Cárdenas's expropriation of the foreign oil companies on March 18, 1938, remains the last major act in the Mexican revolution. This event, which is still celebrated every year as a "day of national dignity," imbues the oil industry with a sense of importance unknown in any other sector of the economy. The 1938 takeover fulfilled one of the salient goals of the revolution—that is, reestablishing national control over subsoil mineral rights claimed by North Americans and Europeans. The petroleum industry ranks with the presidency, Benito Juárez, and the Virgin of Guadalupe as one of the unifying symbols in a nation riven by differences in language, geography, income, education, social class, and political loyalties.

In addition, Petróleos Mexicanos (Pemex) was created in the face of fierce hostility from the expropriated firms, and emerged as Latin America's largest corporation and the Third World's first integrated oil company. Its creation demonstrated that technically competent Mexicans could direct their nation's destiny, contradicting their demeaning stereotype as sombrero-wearing peasants who snore against adobe walls by day only to engage in pistol-waving adventures by night.

Oil also inspired one of Mexico's finest hours—some say its finest hour—when the government successfully resisted foreign attempts to bring the country to its knees following the expropriation of the foreign petroleum holdings. Despite lawsuits, maliciously racist propaganda, and primary and secondary boycotts of its industry, the country retained control of its oil sector. This achievement was all the more remarkable because Mexico's economy was tottering under the triple burdens of depression, reform, and declining food production. In the words of a former director-general of the state monopoly, "Like the Revolution from which they come, the oil expropriation and Pemex are a result of nationalistic drive, faith in ourselves and the affirmation of our sovereignty."[1] Thus it is not surprising that, in a public opinion survey conducted by the *Los Angeles Times* in 1979, Mexico City respondents ranked Pemex second only to the Mexican people as the force likely to exert "the greatest influence over the future of Mexico."[2]

Contributing to Mexico's success in dealing with hostile corporations was the leadership of the indomitable Cárdenas and the decision of Franklin D. Roosevelt's administration to propitiate rather than antagonize Mexico on the eve of World War II. "Perhaps for the first time since the fall of the Empire of Maximilian, the country was able to withstand foreign pressure sufficiently to achieve an important national goal that seriously affected the interests of the great powers."[3]

In response to Cárdenas's pledge that "Mexico will honor her foreign debt," public collections were organized to repay this sudden new obligation. "State governors, high Church officials, patriotic grand dames, peasants, students—all the numberless and picturesque types of Mexicans—pitched in what they had, including money, jewels, even homely domestic objects, chickens, turkeys, and pigs. By August 1938 the collection had grown to about $440,000."[4] The amassing of this amount, though less than 1 percent of the value that the North American firms placed on their properties, reflected the nationalist sentiment aroused by the president's bold decree.

In addition, oil is Mexico's national patrimony; that is, it has served as an inexpensive energy source that has promoted agricultural expansion, generated tens of billions of dollars in foreign exchange, spurred electric output, driven steel production, and ensured other industrial advances. As the head of Pemex told his nation's Chamber of Deputies amid surging petroleum output in the 1970s: "For the first time in its history, Mexico enjoys sufficient wealth to make possible not only the resolution of economic problems facing the country, but also the creation of a new, permanently prosperous country, a rich country where the right to work will be a reality."[5]

The discovery of Mideast-sized oil deposits not only altered domestic conditions, but also profoundly influenced Mexico's international position. Mexican foreign policy changed across a range of substantive issue areas. The very concept of an issue area—as a demarcated arena of policymaking and state interaction underpinned by specific resource endowments—suggests a partial explanation of the observed changes in Mexico's behavior. When the resources available to a nation's policymakers increase or decrease, the nation's ability to influence outcomes within the associated issue area may increase or decrease, and behavioral changes within that issue area are likely to follow.[6] Accordingly, we would expect an expansion of Mexico's oil wealth to modify its behavior in proximate economic matters—for instance, its relations with both the General Agreement on Tariffs and Trade (GATT) and a New International Economic Order (NIEO). But, as the theorists Robert O. Keohane and Joseph S. Nye have stressed, power is not fungible among issue areas, so we have little reason to predict changes in energy-remote matters.[7] Yet, as it turned out, when oil prices increased, Mexican foreign policy changed in a number of arenas where the logic of interaction is unconnected to economic wealth. For example, Mexican policy toward revolutionary movements in Central America was significantly altered. Such policy changes, not driven by shifts in proximate issue-specific resources, can be explained through the concept of "role conception."

The systematic effort to use role types to explain political phenomena profits from the work of Wahlke, Eulau, Buchanan, and Ferguson, who in an extensive study of four state legislatures, discovered that lawmakers developed markedly different orientations toward the norms constituting diverse roles. In relations with constituents, legislators fell into the categories of "trustees," "politicos," or "delegates." Meanwhile, they were classified as "facilitators," "neutrals," or "resisters" with respect to their involvement with pressure groups. The researchers endeavored to examine and explain variations in legislators' role orientations by linking them to ecological, demographic, personality, and institutional factors.[8]

Like legislators, countries often define their distinctive position in regional or world politics in terms of a coherent set of objectives or role conceptions. This idea, as elaborated by international relations theorist K. J. Holsti, embraces "the policy makers' definitions of the general kinds of decisions, commitments, rules, and actions suitable to their state and of the functions their state should perform in a variety of geographic and issue settings."[9] As such, a role conception is an enduring self-image of the appropriate relationship of their state toward the external environment. An easily identifiable role conception is that of Defender of the Faith articulated by President John F. Kennedy when he described the United States as the "keystone" in the arch of freedom and emphasized that it would continue to do its duty as in the past. In his inaugural address, the youthful chief executive stated: "Let every nation know, whether it wishes us well or ill, that we shall pay any price, bear any burden, meet any hardship, support any friend, oppose any foe to assure the survival and success of liberty."[10] Another prominent role is that of Faithful Ally. For instance, in 1967 Premier Pierre Werner indicated that Luxembourg, while "too small to defend itself by its own means . . . has integrated itself with a larger collectivity. Our fidelity to the Atlantic Alliance and our European convictions constitute the base of our foreign policy."[11]

A more assertive role is that of Bastion of Revolution–Liberator, a stance adopted by nations that seek to foment revolution abroad or provide ideological inspiration and material support to foreign revolutionary movements. According to an article in the *Peking Review,* the "victory of China's greatest proletarian cultural revolution not only opened a broad path for consolidating the dictatorship of the proletariat and carrying the socialist revolution to the end, but it has made it possible for China to be a more powerful base for supporting world revolution."[12]

Countries heavily involved in world affairs often see themselves as committed to multiple functions and tasks within the international context. In the 72 countries that Holsti studied, 4.6 was the average number of national roles per state referred to in speeches, press conferences, and other public pronouncements. Yet, the more active states—such as the United States, the Soviet Union, China, and Egypt—saw themselves as playing seven or eight national roles. Among those articulated by U.S. spokesmen were Regional Leader, Regional Protector, Defender of the Faith, Mediator/Integrator, Regional-Subsystem Collaborator, Developer, and Faithful Ally.

Each role springs from both internal factors—such as traditional politics, socioeconomic characteristics, public opinion, and personalities of leaders—and external factors—such as geography, power of neighboring states, alliances, and demand for key exports—and reflects a national consciousness that provides insights into how a country will act when confronted by a given set of circumstances. A state committed to nonalignment avoids military pacts and similar entanglements with Washington and Moscow. However, it is difficult to predict how the nation's other foreign policy actions and its quotidian activities will be affected. Many other roles offer a better opportunity to predict behavior. A government that prizes its status as a Mediator is likely, in the face of a regional or international confrontation, to favor intervention to help resolve the conflict. In the same vein, a government that holds itself out as a Bastion of Revolution will be disposed to thrust itself into any emerging revolutionary episode in the world through diplomatic support, propaganda messages, armaments, and even the dispatch of "volunteers" or regular troops. A Faithful Ally will back its protector's foreign policy goals and, in the case of warfare, fulfill its treaty obligations.[13] States deeply enmeshed in world or regional affairs appear to have more highly structured national roles than those with few and passive role conceptions. The adoption of competing roles may complicate a country's involvement in the international system. The United States' efforts to be a Faithful Ally to Great Britain during the Falklands/Malvinas War collided with its mission—first enunciated in the Monroe Doctrine—as Regional Protector of Latin America. While not explaining every aspect of foreign policy behavior, role conceptions help to illuminate and categorize a country's relationship to the international system.

The thesis of this book is that petroleum prices, expectations about those prices, and the economic conditions arising from changes in oil earnings transformed Mexico's role conceptions in recent years. That is, escalating prices, the expectation of even higher prices, and soaring revenues in the late 1970s and early 1980s catalyzed the pursuit of regional leadership—with pretentions to exercising strong influence within the developing world—in an ideologically charged style that often placed Mexico at odds with the United States over bilateral, regional, and international issues. In this role, Mexican officials stressed their ideological affinity with other developing countries, threw their support behind Nicaragua's Sandinistas and kindred revolutionary movements in the region, spurned membership in the GATT, took issue with the U.S. position on political and economic issues in sympathetic international forums, and endorsed ambitious reforms designed to revise the world economic order by transferring resources from rich to poor states. In short, Mexico sought to restructure international economic and energy relations in order to strengthen the collective bargaining of Third World countries in general and to advance its own interests in particular. On the other hand, anemic domestic savings, military inferiority, and profound dependence on external credits, trade, and tourism constrained Mexico's growing assertiveness. For example, in the late 1970s López Portillo spurned overtures from Venezuela and other cartel members to join OPEC. Nor would he sign an accord with Moscow to ship Mexican oil to Cuba in return for the delivery of Russian crude to a Pemex customer in Western Europe. Transportation savings notwithstanding, such a four-way swap would have raised hackles in Washington. Moreover, at the same time that Mexico trumpeted its close relations with Cuba, it cooperated with the Central Intelligence Agency in efforts to monitor Cuban activity.[14]

Similarly, even as they pursued a vigorous Central American policy, Mexican authorities took pains not to jeopardize economic relations with the United States with which two-thirds of their nation's commerce takes place.

For the symbolic and substantive reasons discussed above, oil is uniquely influential in Mexican leaders' conception of their country's position in world affairs. Almost as impressive as the oil bonanza was the vertiginous rise in agricultural production that took place between 1950 and 1970. Not only did the sharply higher output, which occurred during a period of unprecedented population growth, erase the country's food deficit but also it generated surpluses of corn, wheat, and beans for export.[15] Nevertheless, in contrast to the case with oil, no change in Mexico's international role conception accompanied this highly praised "green revolution."

At the very time when managing abundance seemed to pose the most formidable challenge to Mexico, the appearance of a worldwide oil glut in early 1981 frustrated its quest for regional leadership. The surfeit sparked a protracted decline in prices and slashed Pemex's export earnings by one-third in 1981 alone. Although several months elapsed before Mexican leaders understood the enduring nature of the change, the glut produced a turnabout in the country's role conception. One of the indications of the change was Pemex's substantial shipments to the U.S. Strategic Petroleum Reserve, created in 1975 to safeguard the United States against future acts of blackmail by members of OPEC and other exporters. Mexico's ideologically inspired assertiveness, often linked to mordant reproof of both the international economic order and U.S. policy in the Caribbean basin, generally gave way to pragmatism, moderation, and acceptance of the global distribution of power as Mexico sought a new growth strategy to resuscitate its ailing economy. Important to adopting a new role were the personal qualities of President Miguel de la Madrid Hurtado, who succeeded López Portillo in December 1982, combined with the activism in Central America of the Reagan administration, which severely limited the foreign policy options available to the Mexican government. Largely unaffected by this change was the Foreign Relations Ministry (SRE), which—congruent with its scripted role as a foe of imperialism and champion of Third World solidarity—continued to flail the United States for intervening in Grenada and backing the contra forces in Nicaragua. It is argued here that, the SRE's performance aside, Mexico sought to become a Responsible Debtor —a role characterized by greater reliance on market forces, enhanced cooperation—or, at least, efforts to serve as an Honest Broker—with Washington, and a shift to multilateral negotiations through the Contadora Group to promote its regional interests. The theoretical model in figure 1 illustrates Mexico's transition from Regional Leader to Responsible Debtor, a change, discussed in detail in chapter 2, that is evident in major speeches of Mexican

FIGURE 1
Oil Prices and the Evolution of Mexico's National Role Conception: A Theoretical Model

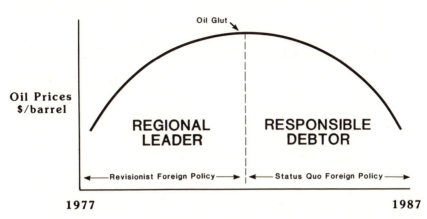

presidents. Such speeches are an invaluable source of information on Mexican role conceptions because, while the legal functions of the country's bicameral legislature appear formidable in international affairs,[16] in reality the chief executive dominates the fashioning of foreign as well as domestic policy. Like his U.S. counterpart, he is constitutionally empowered to control the armed forces, appoint the foreign secretary and other cabinet members, name and remove diplomats, conduct diplomatic negotiations, and make treaties subject to ratification by the senate. In addition, he may declare war pursuant to a congressional resolution, locate and open ports, and establish maritime and frontier customs houses.[17]

Chapter 1 focuses on several of the international roles performed by Mexico during its first 150 years of independence. Chapter 2 analyzes how López Portillo's pursuit of regional leadership gave way, under the impact of the oil glut, to de la Madrid's championing a prudent, cautious, and status quo–oriented foreign policy. Chapters 3 through 7 discuss how the pragmatism of an aspiring Responsible Debtor affected Pemex, Mexico's relations with the United States, its policy toward OPEC, its joint stewardship of an oil-loan pact known as the San José Accord, and its interactions with Nicaragua. The concluding chapter summarizes the findings and discusses the likelihood that, because of inexorable forces, Mexico will become a Contentious Partner of the United States during the administration of Carlos Salinas de Gortari, who will swear the presidential oath on December 1, 1988.

Mexico's National Role Conceptions, 1821–1976

On the spot where the Aztecs once offered throbbing human hearts to appease left-handed Hummingbird, their war god, the Mexican government opened a National Museum of Interventions in September 1981. The facility's seventeen rooms, joined by red-tiled corridors, display photographs, documents, and memorabilia revealing the slights, indignities, incursions, forays, threats, pressures, invasions, and occupations suffered by Mexico at the hands of foreigners since the country declared its independence in 1810.

Mexico has endured various intrusions. The Spanish stayed in the country 300 years following the 1519 arrival of conquistador Hernán Cortés, and Napoleon III dispatched French troops who occupied the nation for five years in the mid-nineteenth century. Nevertheless, the unmistakable focus of the museum is on North American activities in a manner termed "a blend of anti-Americanism and bruised dignity."[1]

The first room prominently displays the Monroe Doctrine, as well as comments of José Manuel Zozaya, Mexico's first ambassador to Washington: "The arrogance of those republicans does not allow them to see us as equals but as inferiors. With time they will become our sworn enemies," the envoy observed. Maps and commentaries describe "Jefferson's expansionism" and the U.S. determination to conquer the American West "at Mexico's expense." Cartoons and engravings recall the 1847 occupation of Mexico by the U.S. Army, led by General Winfield Scott. "U.S. meddling" in the Mexican revolution commands a great deal of space among the displays. Fading brown photographs depict U.S. Marines seizing the flyspecked Gulf port of Veracruz in 1914. Other exhibits brighten the image of Pancho Villa as a revolutionary hero because of General John J. "Black Jack" Pershing's unsuccessful "punitive expedition" to capture the tall, mustachioed combatant following his guerrilla raid on Columbus, New Mexico, in 1916.

This museum is "not a place to stress our losses," stated Gaston García Cantú, director of the National Institute of Anthropology and History and the moving force behind the museum. "No country can afford to lose its historic memory. People must understand what happened and why."[2] Yet it can be argued that the museum reflects the predominantly defensive, reactive, and tutelary foreign policy followed by Mexico since achieving independence in 1821.

In the aftermath of independence, however, Mexican leaders exuded optimism about their nation's position in world affairs. Although warfare had crippled agriculture and the mining industry, the newly minted republic boasted mineral riches that would excite the envy of Crœsus, plus a sizable army, a strategic geographic position between English-speaking North America and Ibero-America, a population almost equal to that of the United States, a land mass that stretched from California to the Isthmus of Tehuantepec, and, later, preferential relations with Spain.[3]

For fifteen years, Mexican diplomats labored toward making their country into a great power, or at least one destined to play the role of Regional Leader. "A series of statesmen—not as numerous nor constant and tightly-knit as the nation needed, but still a number of them—tried to mold a strong and vigorous Mexico that was not afraid to think about power, or about strategic obligations."[4]

Men such as Lucas Alamán, Manuel Eduardo Gorostiza, Miguel Ramos Arizpe, Juan de Azcarate, and José Joaquín de Herrera conducted foreign relations, confident of Mexico's dominant position in international affairs.[5] They sought through the Alamán-proposed "family pact" to implement the Bolivarian vision of a Hispanic America united by history, language, religion, and purpose, but—naturally—under Mexican direction. In this pursuit, Mexico sought to have transferred to Tacubaya, a Mexico City suburb, the 1826 Panama Congress, which drew up a Treaty of Perpetual Union, League, and Confederation that provided for mutual defense, the peaceful settlement of disputes, and renunciation of the slave trade. The scheduled resumption of the Congress at Tacubaya never took place, and only Gran Colombia ratified the treaties signed at the Congress. Even so, Mexican diplomats looked forward to their nation's hegemony over a confederation of Spanish-American countries, which would represent "the most powerful rampart to liberty."[6]

Mexico's aspirations for grandeur prompted it to contemplate the seizure of Cuba from which a Spanish expeditionary force attempted to reconquer Mexico in 1829. These hopes remained unfulfilled, largely because of international rivalries involving the United States, Great Britain, and France, and the opposition of Andrew Jackson's administration to the island's passing to Mexico or another Latin American nation.[7] Authorities in Mexico City also sought influence over the Gulf of Mexico (known as "the Mexican breast"), evinced designs on Puerto Rico, and—despite Guatemala's secession from the United Provinces of Central America, which disbanded in 1938—insisted on a primary role in the strife-torn Central American isthmus.[8]

Domestic turmoil, from the first years of the Republic, dashed Mexican aspirations for a commanding presence in the hemisphere, much less on the global stage. Three centuries of despotic, centralized Spanish governance had emphasized an affinity for authoritarianism, hierarchical rule, personalism, militarism, and a manipulative attitude toward law.[9] The public debt, a large portion of which was assumed from the colonial period, mounted steadily as gold and silver production fell, agricultural and livestock output declined, and wily merchants and importers evaded the import and sales taxes designed to provide national revenue. Exacerbating the situation was Congress's attempt to curry favor for independence by reducing many old taxes, such as those on pulque and tobacco, and eliminating others completely.[10] Not only was the income too low to meet loan installments, but also it was inadequate to cover day-to-day governmental expenses. On the other hand, modest loans from England were helpful to a country whose fiscal structure was patently unsound. Yet "these minor infusions of foreign capital were insufficient to stimulate the economy but did mark the first step of Mexican economic dependency."[11]

Conflict dominated the domestic political arena. The specter of Spanish intervention proved to be the single issue that fostered cooperation between federalists and centralists. The failure of the ill-conceived Spanish attempt at reconquest in 1829 removed this threat and left these groups at loggerheads over both the "rules of the political game" and legitimate institutions for managing conflict and reaching agreement on salient issues. In short, not even the rudimentary aspects of nation-building had attended Mexico's emergence as an independent state.

Mexicans found themselves in a chaotic environment. Bullets not ballots, force of arms not force of persuasion, transferred the reins of government from one caudillo to another. The March 1823 collapse of Colonel Agustín de Iturbide's pretentious nine-month-long "monarchy" ushered in a half-century of kaleidoscopic changes actuated by Machiavellian plots, mutinies, coups d'état, counter-revolts, and civil wars. Between 1823 and 1876—when General Porfirio Díaz installed a dictatorship that endured thirty-five years—Mexicans witnessed the rise and fall of dozens of governments. By 1831 firing squads had taken the lives of four of the five major leaders of the Wars of Independence: Miguel Hidalgo, José María Morelos, Iturbide, and Vicente Guerrero. Eleven of the thirty-six regimes that sprang up between 1833 and 1855 were headed by General Antonio López de Santa Anna, a Mexican Caligula whose name became synonymous with avarice, treachery, deceit, and the dismemberment of the new nation that had once considered itself the great northern cornerstone of a civilization nourished by both Spanish and indigenous currents. In the words of Frank Brandenburg, "Always posing as the savior of his country, [Santa Anna] mutilated, corrupted, and betrayed his nation more than any other Mexican or foreigner before him or since."[12]

The convulsions afflicting internal policy were mirrored in the halls of the chancery as 128 men headed the foreign ministry during Mexico's first twenty-five years of national life. "All concept of continuity vanished; foreign policy was subject to the whims of an extremely agitated political life, and the real meaning of foreign affairs was lost."[13]

This agitated climate facilitated intervention by powers anxious to collect debts, expand geographically in accordance with Manifest Destiny, and carve out a New World empire headed by Old World royalty. Naturally, intervention intensified domestic turmoil. Epitomizing Mexico's degradation and mutilation by foreigners were Spain's occupation of Tampico (1829), Texan Sam Houston's capture of the perfidious Santa Anna (1836), France's victory in the Pastry War waged to collect debts owed to, among others, a French pastry chef (1838), General Scott's seizure of Mexico City (1847), the signing of the Treaty of Guadalupe Hidalgo in which Mexico surrendered a huge portion of its territory to the United States (1848), and Napoleon III's imposition of a Hapsburg, Archduke Maximilian of Austria, as head of the ephemeral Second Mexican Empire (1864).

As portrayed in the National Museum of Interventions, no action left deeper scars than the United States' acquisition of Texas, a part of the northern Mexican state of Chihuahua that in 1836 declared its independence from the federal government in Mexico City. Mexico had recognized the potential political problem posed by the growing population of U.S. settlers and had attempted to limit the influx in 1830. The United States still embraced huge unsettled areas, but poor settlers were hard pressed to pay $1.25 per acre for land that was often inferior to that below the Sabine River, the Texas-U.S. boundary, which sold for one-tenth the price. In 1825, the *Missouri Advocate* declared that the emigration to Texas was explained by the difference between a republic that "gives first-rate land gratis and a republic which will not sell inferior land for what it is worth."[14] Consequently, the illegal flow of U.S. aliens into Texas accelerated, and the newcomers were eager to merge their Lone Star with "the constellation of the stars and stripes."[15]

Texas's declaration of independence in 1836, though not recognized by Mexico, made the new republic a target for intrigue by smugglers, slaveholders, and British and French interests in an atmosphere redolent of burned gunpowder. "Finally, some alarmists feared that the Southern states would gravitate toward Texas and dismember the Union by forming a new confederacy of the South. 'Let us take it [Texas] now,' warned old General Jackson, and 'lock the door against future danger.' "[16]

The controversial 1845 annexation of Texas precipitated a war whose only Mexican heroes were young cadets—the Niños Héroes—one of whom reputedly wrapped himself in the national flag and jumped over the battlements rather than surrender to the despised enemy. Mexicans scorned as patently unjust the Treaty of Guadalupe Hidalgo and the subsequent Gadsden Purchase, in which their country relinquished more than half its territory, including all or parts of what is now Arizona, California, Colorado, Nevada, New Mexico, Texas, and Utah. Compensation under the treaty and purchase totaled only $28,250,000. The long-term costs to their nation of the "War of the North American Invasion," as Mexican textbooks describe the 1846–1848 conflict, continue to shape the attitudes of Mexicans, especially the elite, toward their northern neighbor—a kind of "dormant volcano in the historical conscious-ness."[17] Fatalism suffused Mexican views concerning the year 1848. "From the middle of the sixteenth century Mexican expeditions had been seeking the Gran Quivira in the north, that illusory source of fantastic wealth. And finally it was found, at Sutter's Fort, but a few months too late. The gold of California would not make Mexican fortunes, would not stimulate the Mexican economy, would not pay its share of Mexico's industrial revolution."[18]

Indeed, foreign depredations blemished Mexico's first half-century of inde-pendence. Not surprisingly, the country's international position became "hermetic, nationalistic, distrustful, and defensive; the 'outside' world was only a source of nameless troubles for the country."[19]

Champion of Moral Values

The humiliation of a series of defeats sparked a nationalism that smoldered during both Benito Juárez's liberal administrations and Porfirio Díaz's dictator-ship only to blaze forth in the revolution that began in 1910 and whose goals were crystallized in the 1917 constitution. The treaty signed at Guadalupe Hidalgo also sensitized Mexican leaders to the Hobbesian nature of a world where adversaries prowled, where civilized discourse often failed, where might triumphed over right, and where there was no arbiter to preside over disputed fragments of international law. "From the very beginning of relations among North and South American states, political dialogue was conducted in a vocabulary of union and community," writes Michael O'Leary, but actual policy was based on the normally discredited but still influential politics of international power."[20] Washington's diplomatic and military intrusiveness during the 1910 revolution reinforced this lesson and gave rise to Mexico's support for certain lofty principles as the foundation of its conduct in foreign affairs. The most salient of these tenets were national sovereignty, the juridical equality of nations, national self-determination, and nonintervention in the internal affairs of other states.

In reaction to the U.S. occupation of Veracruz and Pershing's pursuit of Villa, General Venustiano Carranza enunciated Mexico's absolutist position on nonintervention in a doctrine that bears his name: "Any occupation of foreign territory, even when it is inspired by the highest motives, constitutes a hostile invasion and violation of sovereignty."[21] He reaffirmed this policy in another speech: "No country should intervene in any form and for any motive in the internal affairs of other nations. All nations should strictly and without exception respect the universal principle of nonintervention."[22] This emphatic hands-off approach is embedded in the Estrada Doctrine, fashioned by Mexico's foreign minister in 1930, that calls for continuous diplomacy—in other words, the automatic and immediate recognition of a new government, in an act that does not necessarily mean approval of the regime but acknowledges both its existence and political control of its state. In 1939 Mexico modified the doctrine to exclude governments installed with external assistance, so as to justify Mexico's refusal to recognize the Franco regime in Spain.[23]

Mexico has also endorsed two corollaries of the nonintervention principle. The 1868 Calvo Clause asserted that decisions rendered by a nation's courts on the rights and duties of foreigners should be considered final and unappealable to either an alien's home government or his country's diplomatic missions abroad. The clause sprang from a doctrine, also named for Argentine jurist Carlos Calvo, which asserts that no nation has the right to apply diplomatic pressure or employ military force against another nation in order to pursue private claims or collect debts owed to its citizens. Calvo believed that the nineteenth-century European interventions in Latin America were based not on law but on the power of the strong over the weak, thereby violating the canon of the equality of sovereign states. The 1902 Drago Doctrine embraced the international legal principle that no state has the right to use armed intervention, including territorial occupation, to collect public debts. Luis María Drago, Argentina's foreign minister, insisted that such "summary and immediate collection [of a debt] at a given moment, by means of force, would occasion nothing less than the ruin of the weakest nations, and the absorption of their governments . . . by the mighty of the earth."[24] Closely related to Mexico's esteem for nonintervention was fervent support for the peaceful settlement of international conflicts, lest nations like itself possessed of small armed forces suffer militarily at the hands of the powerful.

Clearly Mexico, its postindependence hopes for regional leadership thwarted, turned to the role of Champion of Moral Values in trying to defend itself by emphasizing the precepts of international law. In so doing, it joined other Latin American countries in recognizing its vulnerability to attack by the United States and sought to ensure that disputes could be settled peacefully through recourse to universal principles rather than military means. Ideally, negotiating tables and courtrooms would replace gunboats and battlefields in conflict resolution. Persuading the United States and other powerful countries to respect such tenets would safeguard the unfolding of a revolution in Mexico that, at times, threatened the property rights of foreigners. Its championing of moral values foreshadowed Mexico's emergence as a Defender of Revolutions, a role that became more evident in the 1960s. Ultimately, if other countries adopted these universal criteria, the world might evolve from a dangerous place to one that was safe for weak states. A commitment to broad principles helped explain Mexico's membership in the League of Nations, as well as the United Nations, the International Bank for Reconstruction and Development, the International Monetary Fund, the International Labor Organization, and other specialized agencies of the world organization.[25]

These affiliations notwithstanding, Mexico remained a passive and quiescent member of the family of nations despite the fact that its size, population, and resources could have justified more active participation. President Carranza rebuffed overtures from both the United States and Germany (the infamous Zimmermann telegram) and kept his country "neutral" during World War I. At the close of the conflict, the Allies did not invite Mexico to join the League of Nations because of its pro-German sympathies and activities during the war and because of disputes with Washington over land and petroleum issues. Upon affiliating with the League in 1931, Mexico displayed its traditional concern for resolving disputes through peaceful means. Although far from an activist member of the organization, Mexico condemned German rearmament, backed sanctions against Italy in the wake of Mussolini's attack on Ethiopia, excoriated the Italo-German invasion of Spain, criticized the Anschluss that bound Austria to Germany, and asked that the League take concrete measures against Japan for violating the integrity of China.[26] Antipathy toward the totalitarian states was offset by Mexico's misgivings about the commitment of European democracies to support weak states and defend international law. Denunciation of the dictators aside, Mexico continued to sell oil and other products to Germany, Italy, and Japan, and when the American and British oil companies imposed a boycott, President Cárdenas compensated for losses by signing extensive trade agreements with Germany, elevating oil sales to $2 million by July 1939. Hitler's invasion of the Soviet Union and Japan's bombing of Pearl Harbor solidified Mexican opposition to the Axis powers, precipitating a break in diplomatic relations. The torpedoing by German U-boats of properly identified and fully lighted Pemex tankers in the Caribbean galvanized Mexican support behind a declaration of war on the Axis states. While helpful in supplying workers to the U.S. economy through the *bracero* program, Mexico focused its military conduct on the training of three air squadrons in the United States, only one of which saw action in the Philippines during the last summer of the conflict.[27]

At the conclusion of hostilities, the United Nations attracted enthusiastic

Mexican support because of the promise that it would apply comprehensive tenets of justice to advance the cause of peace and promote the rights of weaker nations in a world dominated by industrial powers. Mexico, one of the fourteen-member Executive Committee that founded the United Nations in San Francisco, submitted several proposals to: expand the powers of the General Assembly, and concomitantly, those of smaller countries; curtail the use of the veto in the Security Council; safeguard human rights; define the rights and duties of states through a binding annex to the charter; incorporate the fundamental precepts of international law into the statutes of member states; and eliminate the concept of domestic jurisdiction as a bar to U.N. action on any matter—whether international or internal in scope—that could constitute or create "a situation likely to provoke international friction."[28] Few of these initiatives were adopted, but they reflected Mexico's acute desire to remake an international system that it found both inhospitable and threatening, thereby reaffirming the country's status as a Champion of Moral Values.

President Miguel Alemán Valdés, who was the second head of state (after Harry S. Truman) to address the General Assembly, welcomed opportunities for Mexican representation on all major U.N. organs, including the Security Council, the Economic and Social Council, and the United Nations Educational, Scientific and Cultural Organization (UNESCO). In his speech, Alemán decried the selfish nationalism of the nineteenth century that had sparked two world conflicts and called on the U.N. to "check the steeds of war" and fortify the foundations of a universal community in which the provocations to war—insecurity, ignorance, poverty, hunger—might forever disappear.[29] Still, Mexico soon displayed apprehension about the U.N. as it became a theater for cold war confrontation. An original nonpermanent member of the Security Council, Mexico did not seek a position in that body after 1946 in order to avoid clashes between Washington and Moscow. Most questions that the council debated did not affect Mexico directly. Nonetheless, on crucial votes Mexico ran the risk of either siding with the Soviets, and thereby alienating the United States, or favoring its neighbor's position and thus appearing as a handmaiden to Uncle Sam. More congenial was UNESCO, which sought to improve the lot of mankind, represent the interests of the underprivileged, and prepare sweeping declarations on human rights. However, Jaime Torres Bodet, Mexico's able chairman of the council, resigned in 1952 when UNESCO's budget was cut.[30] Meanwhile, Mexico showed misgivings concerning what it perceived to be U.S. hegemonic ambitions in the Western Hemisphere. In 1947 Mexico signed the Inter-American Treaty of Mutual Assistance, known as the Rio Pact, to create a system of collective security in the region; yet opposition to possible treaty-inspired interventions in the affairs of other states led Mexico consistently to resist efforts to activate the accord. Although joining the

Organization of American States (OAS), created in 1948, Mexico embarked on a "lonely crusade to prevent the United States from using it as an instrument of its own foreign policy."[31]

Linkage Between Domestic and Foreign Policy

Behind the wall of venerable principles, Mexico concentrated on consolidating its revolution's political, economic, and social gains. This consolidation involved curbing the influence of the armed forces; forging an effective "revolutionary party" that evolved into the PRI; integrating organized labor into the political system; developing a self-sufficient oil industry following the 1938 expropriation of foreign trusts; and formulating a development strategy keyed on erecting protectionist barriers to spur domestic production of goods and services that traditionally had been purchased abroad. Restrictions of imports, inexpensive credit, and subsidies to nascent domestic industries formed the centerpiece of the "stabilizing development" model that impelled income growth exceeding an average of 6 percent annually from World War II until the late 1960s. During this period, Mexico's foreign policy focused on the maintenance of good—or, at least, correct—relations with the United States, which supplied nearly two-thirds of Mexico's imports and provided the chief market for its exports.

Such attention to economic progress at home did not diminish Mexico's devotion to codes of international conduct. This enthusiasm was evident in the aftermath of the 1962 Cuban missile crisis when President Adolfo López Mateos strongly backed the creation of a nuclear-free zone in Latin America. On February 14, 1967, at a ceremony in Tlatelolco, the section of Mexico City where the Foreign Ministry is located, representatives of twenty-one Latin American countries signed the Treaty for the Prohibition of Nuclear Weapons in Latin America. In a joint declaration, the signatories praised what is commonly known as the Treaty of Tlatelolco "as a first step toward global disarmament and ultimately towards complete and universal disarmament." The treaty entered into force on April 22, 1968, following its endorsement by the U.N. General Assembly four months earlier.[32]

By the 1960s, Mexican leaders perceived that a more ambitious foreign policy would be useful in achieving domestic as well as international goals. Political scientist Olga Pellicer de Brody has stressed the domestic implications of Mexico's refusing to join other OAS members in cutting ties with the Castro regime. She avers that this stance, which was extremely popular with the left, enabled López Mateos to conceal economic concessions that he granted to the Mexican right and to U.S. interests.[33] The Cuba policy also enabled Mexico to emphasize its role as a Defender of Revolution, while reiterating its opposition to intervention in the affairs of other states. Burgeoning independence of Washington was displayed in presidential visits to Yugoslavia and Poland and in intensified commercial relations with Moscow.[34]

Nineteen sixty-eight marked a watershed in Mexico's development that, several years later, would affect foreign policy. The stabilizing growth model, highlighted by import substitution, began to fade in the 1960s with the exhaustion of the most promising opportunities for replacing domestically produced capital goods for those manufactured abroad.[35] A slowdown in economic growth directly affected Mexico's pyramidal political system at the apex of which were a strong president and the PRI, victor in every significant election since its founding in 1929. During the twenty-year-plus "economic miracle," the political system furnished opportunities for members of the growing and increasingly heterogeneous middle class. True, they looked askance at the authoritarianism, corruption, and curbs on free expression that marred the regime; yet the system generated both jobs for themselves and social mobility for their children. However, flagging growth exacerbated the essentially political tension that manifested itself in the 1968 student movement. "The majority of middle-class members, intellectuals, students, and business people were altogether relegated to a nonparticipant role, one that simply did not fit with either their social or economic status or their aspirations. This relegated status they shared with labor and peasants, who had also lost in the economic arena."[36] Student dissatisfaction with political conditions and economic inequality sparked at least forty-seven separate demonstrations between July 23 and August 10. In mid-August, the student strike committee shifted from a series of minor protests to a mass rally that drew 150,000 demonstrators to the Zócalo on August 13. There they linked President Díaz Ordaz to police and army violence as indicated in placards that proclaimed: "Criminal," "Hated Beast," and "Assassin."[37] Government concerns about the mounting protests were sharpened by the impending Olympic Games that were to be held in Mexico City and, it was hoped, would portray to the world the image of a developing, innovative, and unified country. The climax to the government-student conflict took place on October 2, 1968, when the army and police units fired on 500 unarmed students, housewives, and office workers who were protesting the lack of freedom in their country.

Echeverría, the interior secretary who supervised the "Tlatelolco massacre," became president in 1970. As chief executive, he proved even more adventurous in using foreign policy to compensate for economic and political immobilism at home. Inaugurated in 1970, he inherited an economy beset by sluggish growth and a political system that had lost legitimacy because of the bloody October 2, 1968, incident. Budget cuts accomplished during his first year in office intensified social protests and led the president to try to rebuild the debilitated consensus that, in the past, had balanced economic justice with economic and social modernization. Therefore, he decided to pursue simultaneously both goals that had produced the post–World War II modus vivendi; specifically, the redistribution of income through a rapid expansion of employment levels and the modernization of an economy burdened with inefficient industries that had grown fat and lazy within the protectionist cocoon.

When private-sector opposition smothered a fiscal reform to expand government income, Echeverría resorted to deficit spending to promote his bold plans for agriculture, mining, education, health, housing, and salary increases. Increasingly, he unveiled his initiatives in populist speeches pronounced in union halls and on communal farms known as *ejidos.* ''The new economic policy came to involve massive state intervention as a means to stimulate the economy and increase the levels of employment. This economic policy was to be complemented by an equally active role for the president in all realms of political life.''[38] Government spending, which was 13.1 percent of GNP in 1970, rose to 39.6 percent in 1976. Meanwhile, the number of state companies grew like Topsy—to exceed 700. To fill posts in new and expanded state agencies, he recruited younger loyalists, many with technocratic rather than political credentials, thereby bypassing a generation of politicians who aspired to rewards for their loyalty to the ''revolutionary'' regime. In the absence of higher taxes, prices rose steadily during the *sexenio,* even as the business community—which intensified its political organizing—suffered from an erosion of profits.[39] Their hostility to his populism prompted Echeverría to decry ''greedy industrialists'' and other ''bad Mexicans.'' The private sector counterattacked, as revealed in a statement by Ricardo Margain Zozaya upon the murder by guerrillas of Eugenio Garza Sada, head of the powerful Monterrey Group:

> It is only possible to act with impunity when the authority has lost all respect; when the State quits being the guard of public order; when not only does the government allow the most negative ideologies to be freely divulged, but it allows them to harvest the negative fruits of hatred, destruction and death . . . when the government has propitiated, through utterances and attacks against the private sector to which the murdered belonged, without any other apparent goal but to promote the confrontation and hatred among the social classes. When no opportunity is missed to foster and help all that is related to the Marxist ideology, knowing that the Mexican people repudiate the system.[40]

Intensifying the headaches suffered by the self-styled ''people's president'' was the Nixon administration's refusal to exclude Mexico from a 10 percent surcharge imposed on all imports in August 1971. This action, which came on the heels of Washington's abandoning the gold standard and devaluating the dollar, gave the lie to the assumption that a "special relationship" with the United States would help Mexico weather any storm that menaced its economy. In response to the import surcharge, Mexico's *jefe máximo* visited Central America and Japan to expand his country's trading vistas in a "multipolar" world.

Echeverría soon shifted from essentially commercial diplomacy to a more aggressive and politicized approach. In particular, he began to devise policies to promote unity between Mexico and other dependent, primary product-producing states and to play a more active role in multilateral bodies. "Echeverría, like other leaders of developing nations, [sought] to use membership in international organizations in a generally unintegrated international system, to try to promote its reconstruction into a more highly integrated and just one."[41] For example, in the U.N., he censured U.S. protectionism, attempted to galvanize Third World solidarity, and launched initiatives to redistribute income from affluent to impecunious states. Moreover, he encouraged the creation of new forums to advance the interests of less developed nations. He was determined to reform the global economic order as a means to facilitate changes at home, which were blocked by entrenched domestic and foreign interests. The portion of the president's annual "State of the Nation" speech devoted to foreign policy demonstrated the mounting salience of this issue area: 1971 (7 percent), 1972 (16 percent), 1973 (15 percent), 1974 (17 percent), 1975 (17 percent), and 1976 (18 percent).[42] Table 1 depicts the vertiginous rise in Mexico's bilateral and multilateral contacts during Echeverría's administration. Underscoring the country's growing reliance on world forums is the increase—from 116 in 1971 to 202 in 1975—in Mexico's participation in meetings of international organizations.

Such meetings were often used to promote Echeverría's basically economic initiatives that included: (1) sponsoring within the United Nations a Charter of the Economic Rights and Duties of States, which would serve as an "alternative to war" between industrialized and less-developed nations by bolstering each nation's sovereignty, while revamping an "unjust system of world exploitation based on both a colonial view of world and the stealing of natural resources and human effort of Third World countries"; (2) designing with Venezuela the Latin American Economic System (SELA), a regional organization created to propel national development by forming OPEC-like cartels in primary products, establishing Latin American transnational companies, safeguarding the prices of manufactured goods and raw materials, and stimulating technical cooperation among developing nations; and (3) creating a Caribbean Multilateral Shipping Company (NAMUCAR), which—like

SELA—would include Cuba but exclude the United States. In addition, he established diplomatic relations with sixty-three countries, the majority of which were underdeveloped African and Asian states; promulgated a 200-mile "exclusive economic zone" to guarantee Mexican control over its coastal resources; enacted legislation to tighten domestic restrictions on foreign investment and technology transfers; and, in concert with most Third World states, backed U.N. resolutions equating Zionism with racism—a position that Mexico repudiated following a highly effective late 1975 boycott of its tourist centers by the North American Jewish community.

Echeverría's international initiatives were not simply a diversionary tactic to shift attention away from internal difficulties; the chief executive "perceived a radical transformation of fundamental aspects of the entire international system as necessary for furnishing solutions to domestic problems and a national malaise."[43] Other factors, like Echeverría's formative political experience, his personal style, and an undisguised ambition to become U.N. secretary-general that bordered on megalomania, helped to account for the president's peripatetic efforts to recodify the world economic order in the name of social justice.[44] Still, domestic considerations loomed largest in his determination to cast Mexico in the role of Regional Leader.

The opportunity for such leadership seemed all the more possible because a number of foreign and domestic factors had attenuated the United States' global influence.[45] Abroad, the Soviets gained nuclear parity and, through the Castro

TABLE 1
Mexico's Major International Contacts During the Echeverría Administration

	1971	1972	1973	1974	1975	1976
Bilateral						
President's visits to foreign countries	1	3	6	11	16	0
Visits by heads of state to Mexico	6	1	2	2	13	4
Countries with which Mexico has maintained diplomatic relations	68	70	75	80	118	131
Treaties and other bilateral instruments signed	5	15	16	27	35	21
Multilateral						
President's visits to international organizations	1	2	1	5	1	1
International conferences held in Mexico	9	5	16	4	9	7
Mexico's participation in meetings of international organizations	116	118	167	143	202	156
Mexican membership in international organizations	73	73	73	74	77	80
Multilateral treaties ratified	3	9	6	9	18	20

Source: José López Portillo, *Quinto informe de gobierno: sector política exterior* (Mexico City: Presidencia de la República, 1981), pp. 111–12.

regime, projected their influence even to the Caribbean Sea, once considered a U.S. lake. Western Europe, united in an economic community, and Japan gradually rebuilt their economies with substantial U.S. assistance after World War II to emerge as dynamic growth centers. Traditional leaders friendly to Washington were replaced in many developing countries, particularly those of Latin America, by technocrats whose prickly sense of nationalism and past exploitation led them to seek autonomy from—and, at times, confrontation with—the United States.

Such new leaders not only helped spawn OPEC, but also made common cause with their counterparts in other parts of the Third World through the U.N. General Assembly, the United Nations Conference on Trade and Development, and the OAS. The Nonaligned Movement emerged as still another venue for challenging the policies of the United States and other Western countries. Defeat in Vietnam proved an especially debilitating experience as the United States lost its first war—an event made even more painful because the victor was a small, underdeveloped communist nation.

On the U.S. domestic front, the reaction to North American destabilization efforts against Allende nourished the post-Vietnam growth of opposition to U.S. intervention in developing countries. Such sentiment limited the options available to the White House, especially the freewheeling deployment of the Central Intelligence Agency to which presidents from Truman to Nixon had become accustomed. In addition, U.S. leaders had to devote much of their attention to energy needs, persistent inflation, cyclical recession, declining productivity, and heightened trade competition.[46]

Echeverría's mercurial behavior, erratic pronouncements, and inept handling of economic problems alienated both the right and left: the former because of resentment toward the government's excessive intrusion into the economy combined with official disdain for the private sector; the latter because the president's performance did not match his rhetoric of reform. Conflicts were subsidized, not resolved. To avoid harsh political and economic choices, the government increased spending—in an indiscriminate and profligate fashion that ballooned the foreign debt from $3 billion in 1970 to $19.6 billion in 1976. The left, whom Echeverría tried to propitiate by condemning undemocratic practices in Spain, lobbying for the removal of OAS sanctions on Castro's Cuba, and championing the cause of the Allende regime in Chile, revealed its hostility in March 1975 when Echeverría visited Mexico's National Autonomous University (UNAM). Traditionally, the president, who has been compared to an Aztec emperor, a Spanish viceroy, a British monarch, and even the pope, enjoys public respect that borders on obsequiousness. But the students defied tradition to hurl epithets, stones, and bottles at the visitor, whom they chased from their campus. Echeverría tried to blame the incident on U.S.

meddling in Mexican affairs. During the disturbance, he branded the ever more abusive student audience as "young fascists, manipulated by the CIA." Later, he told a large political gathering that Mexico "will not tolerate the chain of events which produced the overthrow of President Salvador Allende." No proof was advanced to bolster the charges; however, Foreign Secretary Emilio O. Rabasa declared, "We must assume that the CIA operates in all of Latin America, unless there is proof to the contrary."[47] Throughout his term, Echeverría readily attributed acts of violence to social maladies as varied as drug abuse, casual sex, broken homes, television violence, and yellow journalism.

The crafty populist sought to ingratiate himself with organized labor by requiring thousands of Pemex employees in professional and managerial positions to affiliate with the corrupt, progovernment Revolutionary Union of Oil Workers of the Mexican Republic, now known by its initials as the SRTPRM. He raised industrial wages 17 percent in 1974, 22 percent in 1975, and 16–23 percent in 1976. On October 1, 1976, he also froze the prices of 270 products, including staple consumer goods, whose prices had been allowed to rise slightly after the August devaluation of the peso, the first in twenty-two years. Finally, just twelve days before López Portillo's scheduled inauguration, the chief executive—in an attempt to polish his image as "the people's president"— charged that affluent landowners in the North had broken the law by concealing their holdings under relatives' names. Thus, he ordered that 243,000 acres in Sonora's lush, irrigated Yaqui Valley, valued at approximately $80 million, be turned over to landless campesinos. Government-sponsored unions quickly moved to help the invasion by 8,000 farm families. The seizure stirred a hornet's nest of protest from large farmers and their allies. The Monterrey Group took the lead as businessmen throughout the country closed stores and plants in a twenty-four-hour sympathy strike. Full-page newspaper advertisements flogged Echeverría for "attacking the productive men of Mexico and demanded that the expropriations be canceled." Spokesmen for the private sector warned darkly of the president's attempt to install a "socialist or communist system."[48]

The publicity lavished on the Sonora occupations aroused peasants in neighboring Sinaloa, who threatened to seize 100,000 acres if the government did not continue its expropriations. At that point Echeverría backed down. To prevent an armed battle between peasants and landlords, he announced that only a token 32,000 acres would be handed over to farmhands; additional distribution would await the pleasure of the next president. This statement did not prevent peasants in Durango state from occupying large areas of land. The conflict precipitated widespread rumors that the armed forces would stage a coup d'état. The most prevalent version of this rumor suggested that the military would act to keep Echeverría in power—a ludicrous possibility in view of the president's unpopularity within the officer corps. Four bomb explosions on the eve of the new president's inauguration concluded a sexennium that seemed as absurd as it was tragic.

Not only had Echeverría failed to elevate Mexico to the status of Regional Leader, but also he left his country more dependent than ever on the international economic system in general and the United States in particular. A mounting foreign debt and the need to adopt a stabilization program fashioned by the IMF epitomized this dependency. Until the end of Echeverría's administration, internal values and pressures largely shaped Mexico's international behavior. Clearly, the intervention of the United States and other nations in its affairs explained the country's affinity for moral virtues in world affairs; its own historical experience emboldened Mexico to defend revolutions in Cuba and elsewhere; the quest for regional leadership sprang from an attempt to make the world economic order more congenial to Mexican interests. Yet, the discovery of huge hydrocarbon deposits in Mexico in the mid-1970s meant that the external environment—especially the heightened importance of oil—began to exert an increasingly more powerful influence on both Mexican foreign policy and the nation's role conception.

2

Oil and the Economic Crises

What Echeverría failed to accomplish through populist activism and whirlwind tours—namely, advancing his nation's status as a regional leader—his successor, López Portillo, sought to achieve thanks to the development of huge oil deposits by a semi-industrialized country that is one of the most developed in the Third World. The story of Mexico's second oil boom has been recounted elsewhere and need not be repeated here.[1] Suffice it to say that the discovery of vast hydrocarbon deposits beneath the lush savannas of Tabasco state and in Campeche Sound converted Mexico, a net oil importer between 1968 and 1974, into an ambitiously expansive exporter. In fact, Pemex shipments abroad escalated from 94,438 barrels per day (bpd) when López Portillo swore the presidential oath on December 1, 1976, to 1.5 million bpd when he left office six years later.[2] During this period, announced reserves, buoyed by the addition of prolific offshore fields, shot up from 6.4 billion barrels to more than 60 billion barrels, constituting a faster growth rate than that registered by any other major producer. Consequently, Mexico became the world's fourth largest depository of oil and gas.[3]

This surging output took place amid the strongest seller's market in recent memory, owing to the hard bargaining of OPEC led by Saudi Arabia's Oil Minister Yamani. Moreover, compared to the sheikhdoms, dictatorships, and military regimes that dominated the cartel, Mexico appeared as a stable, geographically secure supplier of a resource whose price rose fourfold in the aftermath of the October 1973 Arab-Israeli War. Pemex exports were particularly attractive to the energy-hungry United States, which had become dependent on imported oil in 1970.

The strong demand for oil gave rise to certain assumptions, which were embedded in both Mexico's Global Development Plan and the National Energy Plan unveiled in 1980.[4] First, obliging international banks, in fierce competition to lend "petrodollars" deposited by Mideast oil exporters, would furnish loans on attractive terms to offset any trade deficit that might exist.

Second, the government believed that it, rather than the marketplace, was best suited to establish prices for Mexico's exports of light Isthmus and heavy

Maya crude. So it would link its prices to those of OPEC, but with a differential added roughly equal to the transport savings enjoyed by U.S. customers by virtue of importing from a nearby Caribbean supplier rather than a Persian Gulf port, from which tankers had to travel thirty-five days to reach U.S. destinations.

Third, Mexico should reduce notably its dependence on the U.S. market, recipient of 83 percent of Pemex exports in 1979, lest oil once more become the target of North American ambition, or if Mideast supplies were again interdicted. As a result, a single customer should neither receive more than half of Mexico's exports nor rely on Pemex for more than 20 percent of its imports. Exceptions were made for Israel and the beneficiaries of the San José Accord.

Fourth, to induce economic diversification, oil should form the adhesive for "package deals" with Japan, France, Canada, the United Kingdom, and other countries. Such agreements "transcended simple and disjointed exchanges by stipulating that receipt of petroleum from Mexico, a dependable exporter that spurned the spot market in favor of long-term contracts, would be contingent upon loans, trade, investments, and technology transfers from these industrialized states.[5]

Above all, Mexican policymakers assumed that the past was prologue to the future; that is, average prices that had risen from $12.57 (1976) to $30.93 (1980) per barrel for Isthmus crude would continue to climb 5 to 7 percent annually—at least until the year 2000. Indeed, López Portillo distributed to his cabinet the Spanish-language version of Jean-Jacques Servan-Schrieber's book, *Le défi mondial*, in which the French commentator stated: "The fact is that oil prices can only continue to rise. After the spiraling upward in ten years from $2 to over $30 a barrel, the price will double again before 1985, over and above any 'gluts' and market forces. And prices will double even sooner should a new 'accident' of the Iranian type occur—something which is quite likely."[6] Hence, the president expected oil exports, whose revenues had soared from $340 million (1976) to $12 billion (1980), to earn enough foreign exchange to expand Pemex operations and to impel economic growth, thereby spurring industrialization and the creation of 12.6 million jobs by 1990. Priorities included manufacturing essential consumer items, stimulating highly productive industries capable of competing abroad, effectively marshaling natural resources, and promoting the output of such capital goods as petrochemicals and steel whose output would leap from 8.3 million tons to 29 million tons by the end of the next decade. In short, more efficient production would spring from the use of cheap energy and from the expansion of both domestic and foreign markets. Industrialization would accompany export diversification so that manufacturers would generate 85 percent of the nation's external income by 1990 when only 15 percent would derive from petroleum. As indicated in figure 2, such unabashed optimism about the oil market and the country's future was associated closely with Mexico's quest for regional leadership.

Mexico as a "Regional Leader"[7]

Mexico's oil wealth proved a magnet for foreign dignitaries. Presidents, prime ministers, princes, and potentates from scores of nations streamed to this cornucopia-shaped nation to seek access to its newfound energy wealth. Although Echeverría's move to expand his country's influence in the area failed as economic difficulties intensified Mexico's need for financial assistance from the IMF and international banking sources, hydrocarbon riches enabled his successor, a former finance secretary, to pick and choose from among dozens of offers to establish bonds with other industrialized countries, as well as those in the developing world. "The demonstrated capacity to back rhetoric with resources differentiated Mexican foreign policy under López Portillo from earlier phases and marked the country's transformation from a minor power to a regional one."[8] As depicted in table 2, Mexico demonstrated its international élan through increasing international contacts, especially during López Portillo's first four years in office.

President Dawda K. Jawara's October 1982 visit to Mexico City made Gambia the thirty-ninth country to send a high-level delegation in search of energy resources. "For the first time, we have options," stated Carlos Martínez Ulloa, a Columbia University Ph.D. who served as the nation's director of public debt.[9] A Latin American diplomat expressed this sentiment more poetically: "With the discovery of oil, Mexico has replaced Brazil as *la niña bonita*, the 'pretty girl,' courted by all the rich suitors in town."[10]

Government officials were not the *niña bonita*'s only admirers. In the mid-1970s, bankers scorned the Echeverría regime as an economic pariah, to which only short-term, high-interest loans could be contemplated. However, following revelation of the petroleum bonanza, they competed fiercely to finance Mexico's oil-powered development. "Planes couldn't get us there fast enough to lend them money," confided a senior vice president of a New York bank.[11] When López Portillo visited Japan, banks were practically throwing money at him as a syndicate headed by the Bank of Tokyo, Mitsubishi Bank, and the Industrial Bank of Japan made loans totaling $600 million. A measure of Mexico's growing influence appeared in its ability to attract ever larger loans on increasingly favorable terms—that is, lower interest rates, longer repayment requirements, and generous grace periods. For instance, in 1976 Mexico had to pay 1.75 points over the London Inter-Bank Offering Rate (LIBOR), the benchmark figure that international banks charge one another. Three years

FIGURE 2
Oil Prices and the Evolution of Mexico's National Role Conception: From Regional Leader to Responsible Debtor

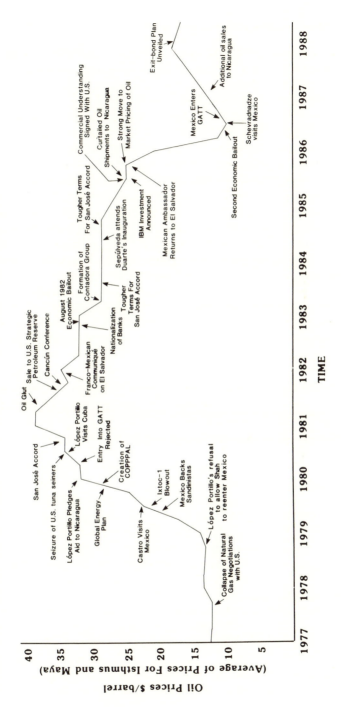

later, Mexico could secure ten-year loans for .625 of a point above LIBOR. The ease in obtaining credits, the lion's share of which came from U.S. financial institutions, helped to swell the public sector's external debt from $22.9 billion in 1977 to $85 billion by 1982.

Mexico's abundance of hydrocarbons prompted foreign leaders to unfurl the red carpet for López Portillo. All told, he accepted invitations to visit eighteen countries, meeting with such prominent statesmen as President Valéry Giscard d'Estaing, Premier Pierre Elliott Trudeau, Premier Masayoshi Ohira, and Chancellor Helmut Schmidt. The IMF also recognized Mexico's standing as a growing oil producer. The organization had imposed a $3 billion limit on new foreign borrowing by the Mexican government in 1977—a figure that was to decline to $2 billion in 1978. The IMF's willingness to extend the $3 billion cap to 1978 permitted the country access to another $1 billion in credits used largely by Pemex to purchase oilfield equipment. Impressive earnings from petroleum sales, success in meeting IMF targets, and the timely repayment of loans ensured the removal of all restrictions on external indebtedness.[12]

An analysis of the annual State of the Nation speeches reveals that, with every increase in oil prices and reserves, López Portillo more boldly advanced his country's bid for regional leadership. As he insisted in his second *informe* to the nation: "We are participating with increasing determination in the effort to transform the international system. We want a tomorrow, if insanity does not prevent its attainment, in which the world is one of co-responsible nations and

TABLE 2
Mexico's Major International Contacts During the López Portillo Administration

	1977	1978	1979	1980	1981	1982
Bilateral						
President's visits to foreign countries	3	6	2	8	2	2
Visits by heads of state to Mexico	6	3	10	6	15	7
Countries with which Mexico has maintained diplomatic relations	134	134	138	139	140	141
Treaties and other bilateral instruments signed	33	50	47	118	26	20
Multilateral						
President's visits to international organizations	1	1	1	0	0	1
International conferences held in Mexico	10	22	20	20	68	20
Mexico's participation in meetings of international organizations	267	271	286	275	623	434
Mexican membership in international organizations	81	82	83	83	83	83
Multilateral treaties ratified	16	13	8	9	15	27

Source: Miguel de la Madrid H., *Primer informe de gobierno: sector política exterior* (Mexico City: Presidencia de la República, 1983), pp. 100–01.

not of superpowers."[13] The president's audacity became evident in Mexico's bilateral relations with the United States, in his country's defense of revolutionary movements in the Caribbean basin, and in activism in international forums on behalf of restructuring the global economic order.

Mexico City and Washington crossed swords frequently during López Portillo's presidency. The first serious clash occurred late in 1977 when the Carter administration refused to permit Border Gas, Inc., a consortium of six pipeline companies, to purchase 2 billion cubic feet per day of natural gas from Pemex at a price linked to No. 2 fuel oil, a comparable energy source, delivered in New York harbor. This refusal, which sprang from executive-legislative political considerations in the United States, infuriated the Mexican chief executive, who claimed to have been left "hanging by his paintbrush" when U.S. officials knocked the ladder from under him. The contretemps preceded more than a year and a half of mutual recriminations before negotiators from both countries reached an accord on gas exports, albeit at a lower level than originally proposed.[14]

Sparks also flew over other issues. Especially vexing to Washington was López Portillo's unwillingness to consider compensating resort operators, property holders, and other Texans who suffered damages because of Ixtoc-1, an offshore Pemex well whose June 1979 "blowout" spewed a torrent of hydrocarbons into the Gulf of Mexico. Jorge Díaz Serrano, director-general of Pemex, stated that "there is no international regulation for this type of accident. Therefore, we must abide by Mexican laws and legal principles."[15] Other acts of assertiveness toward the neighboring superpower included Mexico's refusal to allow the shah of Iran to return to his Cuernavaca home following medical treatment in New York in late 1979, the seizure and fining of U.S. tuna seiners for violating its 200-mile economic zone (a move that automatically triggered a U.S. embargo on Mexico's tuna and tuna products), and rejection of a North American Accord, proposed in the 1980 campaign by candidate Ronald Reagan to foster U.S.-Mexican-Canadian economic integration.[16] "We want the United States to treat us as a mature country capable of managing our own affairs"—was the theme articulated by a host of Mexican officials. As evidence of the greater oil-inspired confidence exhibited by Mexico, Mario Ojeda has cited the differences in the first and second meetings between Presidents López Portillo and Carter.[17] At their February 1977 session, the Mexican leader plaintively emphasized the importance of moral responsibility in bilateral relations. "The United States has to face up to a great responsibility in regard to Mexico," he said, "not only as a geographic neighbor, but also as a neighbor which finds itself in a process of development."[18] Two years later in a luncheon toast to Carter, the Mexican chief executive exhibited greater self-assurance, as well as disappointment over the natural gas issue, when he stated:

Our peoples desire agreements in depth and not circumstantial concessions. Between permanent and not merely occasional neighbors, sudden
deception and abuse are poisonous fruits which sooner or later turn back
upon one. It is within this perspective that the complex phenomenon of
our interrelationship must be situated, an interrelationship which must
not under any circumstances be confused with dependence, integration,
or dilution of frontiers. The two countries complement and need one
another reciprocally, but neither would desire to depend on the other to
the point where its international actions were annulled or where the space
for its international action was reduced, or where it lost its self-respect.[19]

Oil riches strengthened Mexico's determination, first manifest in the 1960s,
to defend regional revolutions. López Portillo, who was much more hospitable
to Fidel Castro than to Jimmy Carter, moved Mexico closer to Cuba than ever
before in the history of the two nations. Visits, accolades, and bilateral
protocols nourished this relationship. In May 1979 Castro returned to Mexico
for the first time since setting forth from Tuxpan twenty-three years before to
capsize the Batista dictatorship. Amid gun salvos, mariachi bands, and a
cheering crowd, López Portillo lauded his guest as "one of the personalities of
the century."[20] In sharp contrast to the mordant toast that he inflicted on the
U.S. president three months before, the Mexican leader commended the
"fraternal and indestructible" friendship between Mexico and Cuba; deplored
Washington's economic embargo of the revolutionary regime; decried the
continued existence of the Guantanamo Bay naval facility as an imperialist
enclave; and scorned the U.S. contention that Cuba played a key role in the
success of Nicaragua's Sandinistas.[21]

In turn, Castro lauded Mexico, its best ally in the Western Hemisphere, as
the only Latin American nation to maintain diplomatic relations with his regime
during twenty years of "immense obstacles, misunderstandings, isolation, and
blockades." He also praised Mexico for defining its energy policy "not in
terms of the oil needs of the United States but as an instrument for the
development of your own country." In an oblique reference to Carter's recent
visit, the Cuban president noted that he had not "come to seek Mexican oil or
gas which seems to be fashionable these days." He also urged "just, civilized,
and humane treatment" for Mexico's undocumented workers in the United
States. Such migration, he insisted, was "the bitter and inevitable fruit of the
mutilation of Mexican territory and the underdevelopment imposed by the
force, arrogance, and domination of the United States in the past."[22] The visit
spawned an agreement covering economic cooperation and technical-scientific
cooperation in the production of sugar and its derivatives.

López Portillo's mid-1980 visit to Cuba showed solidarity with Castro's regime at a time of deepening economic problems, which were illuminated by an exodus of 125,000 refugees, or "Marielitas," to the United States. An estimated crowd of one million turned out to hear Castro extol Mexico's steadfastness as an ally. In one of his speeches during the visit, López Portillo insisted: "We will allow nothing to be done against Cuba." To him, acts against Cuba were tantamount to acts "against ourselves."[23] The meeting paved the way for a far-reaching accord on energy development begun the following December. And in late 1981, when Castro desperately needed hard currency to make debt payments to Western banks, the Mexican government secretly loaned Cuba $100 million.[24]

The Mexican president's receptiveness to revolutionary movements in Central America further underlined his country's emergence as a Regional Leader anxious to carve out its own sphere of influence. Backing these groups enabled Mexican leaders to demonstrate their independence of the United States. Meanwhile, even while following relatively conservative policies at home, government officials could declaim their continued support for revolution abroad in hopes of winning favor in leftist and nationalistic party, academic, intellectual, labor, and media circles. The symbolic aspects of its foreign policy must not blind us to the fact that Mexico has viewed Central America through a North-South optic compared to the East-West perspective of the Reagan administration. Briefly stated, Mexican leaders, whose views should not be considered as "momentary flashes of anti-Americanism or sudden surges of machismo,"[25] believed that the regional crisis sprang not from communist intervention but from poverty, unemployment, political repression, social injustice, and a highly skewed pattern of landownership. Thus, U.S. military intervention against broad-based opposition movements would not only fail; it would deracinate hundreds of thousands of people who might flood into the impoverished states of southern Mexico. Even worse, the United States would have an armed presence on both Mexico's northern and southern flanks, making the country's precious oil fields even more vulnerable to seizure in the event of another international energy crisis. As Bruce Michael Bagley has argued, " 'Regionalizing' the conflict in Central America would force the nation's civilian elites to grant larger budgets and greater decision-making power to the armed forces."[26] Thus, for the sake of stability the Mexicans considered it prudent to befriend revolutionary movements in the hope of becoming an important source of economic and technological support, moderating their development, and thereby purchasing an "insurance policy" against the export of revolution.

Consequently, the Mexicans preferred to harness and deflect revolutionary forces in neighboring countries rather than attempt to quash such aspirations to preserve the status quo. Through continuous contacts, both political and economic, the Mexicans hoped to moderate revolutionary change in their region. Thus, López Portillo believed it imperative to foster ties with leftist movements that were destined to topple conservative regimes, in hopes of shaping their agendas and obtaining their friendship for his country. After all, a pluralistic, single-party-dominant government chosen through elections developed in Mexico following a revolution, and the rise of "left-wing PRIs," would be infinitely preferable to either right-wing dictatorships or protracted superpower-aided warfare. Also prevalent in Mexico City was the view that "despite their Marxist-Leninist trappings, Latin American revolutionaries continued to be middle-class nationalists with a popular following."[27] Mexico could cite its cordial relations with Cuba, which had not fomented insurgencies on Mexican territory, as proof that cooptation worked.[28]

Under López Portillo, Mexico became especially supportive of Nicaraguan revolutionaries in the Sandinista Front of National Liberation (FSLN). Following an abortive FSLN-led popular uprising against General Anastasio Somoza Debayle in September 1978, Mexico's Managua embassy, a focal point of anti-Somoza conspiracies, welcomed hundreds of political dissidents in a marked departure from the country's nonintervention policy. Moreover, Gustavo Iruegas, who was then in charge of Mexico's embassy in Managua, allegedly aided the rebels by setting up "safe houses." In addition, the Mexican government and the PRI appropriated funds for the Sandinista rebels and their political front, the Group of Twelve, composed of well-regarded professors, priests, and businessmen.[29] The following spring, López Portillo denounced the "horrendous genocide" in Nicaragua and severed diplomatic relations with Somoza, justifying the rupture on the grounds that Somoza's repressiveness of the civilian population "transcended internal affairs and affected the other nations of the area, threatening the stability and peace of the region.[30] Mexico, which advocated the dictator's replacement by the Sandinista-led "government-in-exile" that operated from San José, Costa Rica, dispatched missions to Venezuela, Colombia, Panama, Jamaica, and the Dominican Republic to solicit withdrawal of their ambassadors from Managua. Meanwhile, Pemex terminated oil sales to the Somoza regime.

At a June 1979 OAS meeting, Mexico's foreign minister defended "the sacred right [of the Nicaraguan people] to rebel against tyranny" and spearheaded efforts to thwart a Washington-sponsored plan to dispatch an inter-American peace force to Nicaragua to restore order, prevent the Sandinistas from taking control, and supervise free elections. This issue symbolized both Washington's diminished influence in the region and the strong opposition of Latin American governments to the concept of a peacekeeping detachment. Fourteen years earlier, the Organization of American States had pliantly endorsed the occupation of the Dominican Republic by U.S. forces. In 1979 it turned thumbs down on similar intervention in Central America, even after Washington cited alleged Cuban involvement in behalf of the FSLN guerrillas. Cuba was no longer the outcast it had once been—as fourteen Latin American countries boasted diplomatic ties with the Castro regime. López Portillo sent ammunition to the Sandinista southern front, and on July 19, two days after Somoza fled, the Mexican presidential airplane carried the newly formed junta from Costa Rica to Nicaragua.[31] In López Portillo's view, "Everything concerning the political organization of the Nicaraguan people is an internal matter that only the Nicaraguans can resolve."[32] He moved swiftly to fill the influence vacuum in the region left in recent years both by the United States' preoccupation with Europe, the Soviet Union, and Asia, and by Cuban adventurism in Africa. For example, the Mexican president showered praise and aid on the new Government of National Revolution (GNR)—even though Washington provided more aid than any other nation during the fledgling regime's first year and a half in power.

On January 24, 1980, the Mexican leader flew to Managua for a nine-hour visit with the country's new leaders. In a speech richly colored with the Cervantesque quality esteemed in the world of Hispanic letters and politics, he called their revolution a path that other countries could follow to escape Latin America's "pitiful labyrinth"—just as the Mexican and Cuban revolutions had served as models in earlier decades. In an unmistakable allusion to Somozan loyalists and the United States, he urged nations of the hemisphere to fight "against internal demons and the satanic ambition of imperial interests." Praise of the revolution presaged an outpouring of Mexican diplomatic, technical, and economic support as López Portillo expressed his government's intention to aid Nicaragua "without conditions of any kind, as Mexico has been doing to date." The joint communiqué issued at the conclusion of the visit stated that bilateral relations were "based primarily on the traditional brotherhood of their peoples and reaffirmed in the present era by mutual understanding and support, as behooves countries which have waged revolutionary struggles for the benefit of their peoples."[33] The document also cited appropriate areas for Mexican-Nicaraguan cooperation—notably, mining, oil, fishing, construction, shipbuilding, health, education, transportation, and public administration. Several specific agreements were reached at this meeting as well. Mexico offered technical assistance and agreed to serve as an intermediary between the GNR and the international financial community. Mexico also promised to furnish 315 tons of equipment to reconstruct Nicaraguan telephone and telegraph lines. In addition, López Portillo granted credits for the purchase of medicines, medical supplies, and raw materials. He also turned over a Mexican-owned fertilizer plant, offered monies to facilitate the purchase of 150 Mexican buses, and excused repayment of $17 million in loans contracted by the Somozas. Privately, Central Bank of Mexico executives claimed that as much as $200 million in cash may have been handed over to the *comandantes*. Yet the most ambitious aid came in the form of oil, the subject of chapter 7.

In October 1979, Mexico—through the PRI—hosted twenty-two "nationalistic, democratic, and anti-imperialist" parties from fourteen nations in Oaxaca where they endorsed "revolutionary nationalism" as the guiding philosophy of a Permanent Conference of Latin American Political Parties (COPPPAL).[34] The new grouping included parties of different ideological hues—namely, socialists such as Costa Rica's National Liberation party; populists such as the Brazilian Workers' party; Eurocommunist-oriented parties such as Venezuela's Movement Toward Socialism; and the Sandinistas, who referred to themselves as a people's political-military front. While welcoming Social Democratic groups, the organization rejected Communists and Christian Democrats on the grounds that they were "under foreign influence"—a specious conclusion in the case of the latter. The result was to divide Christian Democrats and Social Democrats, which weakened two of the most important democratic groupings in the hemisphere.[35]

COPPPAL committed itself to the vague and noncontroversial goals of joint action, mutual support, open communications, and formal relations among its members. In fact, at its founding session and subsequent gatherings, the organization furnished a forum for members to denounce multinational corporations, the International Monetary Fund, and the United States and other developed countries. Communiqués issued at the conclusion of COPPPAL meetings often appeared as disjointed compilations of grievances zeroing in on external forces deemed manipulative and imperialistic.

At first glance, COPPPAL seemed to have advanced Mexican goals by enhancing the country's stature, giving a rhetorical fillip to López Portillo's Global Energy Plan, and bringing the PRI into contact with a number of small, left-wing movements that someday might prove to be as significant a force in their countries as the Sandinistas have been in Nicaragua. Above all, Mexico sought through the newly formed body to domesticate the Sandinistas by surrounding them with moderate parties that might temper their revolutionary zeal. "We want the Nicaraguans to feel they can look to us if they need anything," explained one Mexican diplomat. "They don't have to think the Cubans are their only friends."[36] Although designed to further Mexican interests and invariably headed by the president of PRI, COPPPAL at times appeared to be the proverbial tail that wags the dog. Specifically, the ease of including items in the organization's declarations found Mexico identified with radical positions that were at odds with its own national goals.

Assertiveness also characterized López Portillo's Salvadoran policy. In early 1980, the PRI president met with commanders of the insurgent Faribundo

Martí National Liberation Front (FMLN) and their political allies in the Democratic Revolutionary Front (FDR). In August, Mexico reportedly toyed with the idea of recognizing the FDR, with which the Foreign Ministry had established high-level and continual communications and which was seeking to overturn the U.S.-supported civilian-ministry junta.[37] A few months later, Mexico warned Washington against intervening in El Salvador, describing the country as "a zone of hegemonic confrontation" which had been "elevated to the undesirable rank of a strategic frontier." These warnings were references to the U.S. argument that the Salvadoran conflict could impinge on North American national security.[38]

In September Mexico and the recently elected government of François Mitterand submitted a joint communiqué to the U.N. Security Council, urging that the insurgent FMLN and FDR be recognized as a legitimate and "representative political force." As such, it "should participate in the establishment of mechanisms of rapprochement and negotiation necessary for a political solution to the crisis in El Salvador."[39] This declaration, which the insurgents and their political allies greeted enthusiastically because it came during a period of military weakness, sparked denunciations of "interference in El Salvador's internal affairs from a dozen Latin American and Caribbean nations. Still, the statement played an important role in generating support for the FDR-FMLN from Western European countries as registered in U.N. votes on resolutions affecting El Salvador.[40]

Also reflective of its growing influence in the area was Mexico's emergence as a significant aid donor through the provision of discounted oil and low-interest loans to ten countries of the Caribbean basin. The San José Accord, the vehicle for this "unselfish international cooperation," was undertaken with Venezuela in August 1980. This oil facility, which is the focus of chapter 6, marked the first time that two Third World nations had embarked upon a joint aid program to benefit other developing states.

Mexico's prominence as a growing oil exporter precipitated more active peacemaking efforts. López Portillo successfully continued to canvass for signatories to the Tlatelolco Treaty, gaining the adhesion to Protocol II of the Soviet Union (1978) and France (1979), among others. The Mexican president also characterized his country as an international peacemaker in a September 27, 1979, speech to the General Assembly in which he unveiled a Global Energy Plan.[41] Just as his predecessors had stressed a nuclear-free Latin America as an important step toward world peace, López Portillo's goal was to fashion a framework for the pacific resolution of energy-related problems, thereby avoiding a possible war between consumers and producers in the wake of a stunning rise in oil charges. Like Echeverría before him, he believed that international harmony could be achieved only by restructuring economic relations in a manner congenial to the interests of developing states. Yet, as will be discussed in chapter 5, his program came a cropper because of unyielding hostility from OPEC countries. Still, this opposition failed to halt the ambitious chief executive's missionary work on behalf of what turned out to be a stillborn project. Nor did it prevent his nation's winning a nonpermanent seat on the Security Council in January 1980 as a compromise candidate after neither Cuba nor Colombia could muster the necessary two-thirds majority in the General Assembly. Oil-induced confidence emboldened Mexico to promote initiatives in the U.N., especially with respect to energy, human rights, transnational corporations, global development issues, the Law of the Sea, and migration.

The migration question quintessentially illustrated López Portillo's attempt to shift to an international forum a bilateral question involving the United States. In reference to his government's inaction with respect to the accelerating flow of illegal workers across the Rio Grande, he stressed that "Mexico cannot constitutionally, politically, juridically, or morally agree to undertake repressive measures to impede the movement of its citizens inside or outside its territory." Thus, to deal with the "undocumented worker problem," he proposed that the United Nations adopt a "Code of Conduct on the Rights of Migratory Workers."[42] Washington was unwilling to turn over to an international agency a traditional and widely recognized element of sovereignty: control of a nation's borders.

Mexico dispatched representatives to other multilateral forums, as indicated by the fact that the number of international conferences in which it participated shot up from 156 in 1976 to 623 in 1981. The Nonaligned Movement acknowledged Mexico's regional leadership by asking Foreign Secretary Castañeda to address its sixth conference, held in Havana in September 1979. Even more important, López Portillo assumed along with Austria's Chancellor Bruno Kreisky the chairmanship of the International Meeting on Cooperation and Development that convened in the Mexican resort of Cancún on October 22, 1981.[43] The purpose of this twenty-two-nation summit was to lay the groundwork for global negotiations that Mexico and other Third World states hoped would commence a large-scale resource and technology transfer from developed countries of the North to developing states of the South. Although nothing tangible emerged from the conclave, López Portillo stated that the "spirit of Cancún" had influenced the seven most industrialized nations of the West to endorse global negotiations at the 1979 Versailles summit.[44] When asked if Mexico's zeal in behalf of the impecunious and dependent would affect relations with Washington, López Portillo said: "Mexico has its own foreign policy. It takes its decisions independently and it expressed them in a world from which, fortunately, Mexico's voice is being heard and will be heard more and more." That the Group of 77, composed of less-developed nations, selected Porfirio Muñoz Ledo, Mexico's U.N. ambassador, as its president in 1983 provided additional evidence of Mexico's continued activism in arenas dedicated to restructuring international power relations.[45]

Petrolization and the Oil Glut

Immediately upon taking office, López Portillo moved to close the chasm between the state and the private sector that had opened during Echeverría's *sexenio*. To do so, he launched an "Alliance for Production" composed of the government, industry, and organized labor. In return for the government's promise to put its house in order, recognize the essential role of the private sector in a mixed economy, and provide economic incentives, the business community pledged to operate more efficiently and expand investment. For its part, the union movement, which would benefit from both the new jobs generated by investment and the vigorous attack on inflation, agreed to hold demands for wage increases to 10 percent (later 13 percent) and foster harmony in the work place. Labor's assurances were made by Fidel Velázquez, the redoubtable head of the Confederation of Mexican Workers (CTM), the nation's 4.5 million–member labor confederation and an integral part of the PRI.

The rhetoric of alliance, cooperation, and collaboration obscured the presence in the cabinet of two factions: the "nationalist-populists" and the "liberal-rationalists." The former, associated with the so-called Cambridge school of economics and epitomized by Secretaries Carlos Tello (Planning and Budget) and José Andrés Oteyza (Patrimony), favored ubiquitous state intervention in a relatively closed economy, a subordinate public sector, widespread subsidies to win labor and peasant support, and iron constraints on foreign investment to guarantee Mexico's independence from external control. The latter, represented by Finance Secretary Julio Rodolfo Moctezuma Cid, believed in stimulating market forces in an open economy that boasted a robust private sector, a pragmatically defined status for foreign investment, increased trade, fiscal conservatism, and a government that concentrated on productive investment in infrastructure rather than on competition with the business community.[46] Differences between the two groups erupted over the budget for fiscal year 1978, prompting the president to dismiss both Tello and Moctezuma Cid. Nevertheless, this conflict was soon forgotten as the dollars that poured into Mexico for black gold expanded the GDP by approximately 8 percent each year between 1978 and 1982. This growth generated vast profits for the private sector and helped create one million new jobs annually.

Mexico's success became the envy of a world floundering in recession. Such glamor, however, diverted attention from the beginnings of "petrolization"—a neologism connoting an overheated economy fueled by oil revenues, an over-valued currency, mounting dependence on external credits to import escalating amounts of food, capital, and luxury goods (which rose from $6 billion in 1977 to $23 billion in 1981), a stagnant agricultural sector, and—above all—outsized budget deficits spawned by prodigious public spending. Compounding the problem in Mexico was the shortage of skilled workers amid widespread unemployment, as well as bottlenecks in port services, storage, and transport. Highways had been built to handle one-fifth of the traffic of the 1980s,[47] while the railroads had been little changed since the days that Pancho Villa and his troops rode them during the revolution. Rather than raising taxes, Mexican leaders chose to cover budget shortfalls by printing stacks of crisp new peso notes. As one specialist on Mexican affairs described the situation: "The oil boom had aroused exaggerated expectations in all segments of Mexican society. The government was tempted to grant all of the demands, for expanded social programs, heavily subsidized consumer goods and services produced by the public sector (such as cheap gasoline and electricity), lavish subsidies to private-sector producers (such as increased export subsidies, reduced import duties), large-scale infrastructure projects and other capital investments."[48]

Excessive spending drove prices ever higher. The rate of inflation dropped from 20.7 percent in 1977 to 16.2 percent in 1978 only to climb back to 28 percent in 1980. A dearer peso vis-à-vis the dollar discouraged tourism, inhibited export of relatively labor-intensive manufactures, and deepened dependence on oil and its derivatives to generate dollars. Hydrocarbons, which accounted for 21.9 percent of the nation's export earnings in 1977, produced three-fourths of these revenues six years later. (See table 3.)

The collapse of the shah's regime and the outbreak of warfare between Iran and Iraq sparked a tripling of oil prices. Mexico followed OPEC "hawks" to widen substantially the difference betwen its Isthmus export price and that of Arabian Light, OPEC's marker or reference crude. Confident of the security of supply, buyers paid as much as a $6.50 premium for Mexico's best crude, the charge for which soared to $38.50 per barrel. The spiraling prices, combined with Mexico's brisk growth rate, blinded many observers to the danger signals flashed by the Mexican economy. Meanwhile, bankers kept arriving with their briefcases bulging with applications for loans, whose interest rates expanded along with Mexico's indebtedness.

Like a heroin addict who sells his blood in the morning to get a "fix" from an eager, well-heeled supplier at night, Mexico coped with the pressures of petrolization by exchanging oil for loans. The situation changed dramatically after an international oil glut appeared in the spring of 1981. A medley of factors generated the 2 to 3 million bpd surfeit—namely, Saudi Arabia's record output of 10.3 million bpd, energy conservation in the face of unparalleled crude prices, a worldwide recession, and expanded production in Mexico, Great Britain, the Soviet Union, Alaska, and other non-OPEC areas.

Determined not to "rat on OPEC," Mexico tenaciously adhered to its prices—$38.50 per barrel for Isthmus; $32 for Maya—even as clients abandoned Pemex either for exporters offering discounts or the spot market. Disquieting changes occurred within an eight-week period beginning June 3, 1981: Mexico cut its prices $4 a barrel; Díaz Serrano, Pemex's director-general and architect of the price reduction, abruptly resigned under fire; Patrimony Secretary Oteyza, an inveterate nationalist-populist, assumed control of oil policy; and Pemex shipped 109.15 million barrels to the U.S. Strategic Petroleum Reserve. When the smoke cleared, Mexico was selling a mixture of Isthmus and Maya at $30.70 per barrel, only 10 cents above the price that ushered in the chaos. For the year, Pemex earned but $14.6 billion from oil, gas, and petrochemical sales, just two-thirds of the amount projected.

The tumult of 1981 excited misgivings about the competence of López Portillo's administration, now strongly influenced by nationalist-populists. This apprehension manifested itself in a quickening flight of capital, which accelerated even more when the government—chary of alienating organized labor, the PRI's sturdiest pillar, with presidential and congressional elections scheduled for midyear—declined to impose rigorous stabilization after the February 1982 floating (and 31 percent devaluation) of the peso.

Even after de la Madrid's overwhelming victory in the July 4, 1982, presidential election, the government failed to act decisively. By early August,

TABLE 3
Mexican Oil Production, Exports, and Export Earnings, 1977–1987

	Crude Production (thousands of bpd)	Crude Exports (thousands of bpd)	Total Hydrocarbon Exports ($ million)	% of Total Exports
1977	981	202	1,018.7	21.9
1978	1,209	365	1,837.2	30.3
1979	1,461	533	3,986.5	44.7
1980	1,936	828	10,401.9	68.0
1981	2,312	1,098	14,574.0	75.0
1982	2,746	1,492	16,594.5	79.0
1983	2,665	1,537	16,164.9	74.9
1984	2,685	1,525	16,601.3	69.0
1985	2,630	1,438	14,606.0	68.0
1986	2,428	1,290	6,132.0	38.9
1987*	2,604	1,237	4,211.1	40.6

Source: "Economic Trends Report," U.S. Embassy, Mexico City, February 1987, p. 18.
*January–June.

reality began seeping into Mexico's economic fantasyland. At the beginning of the month, the government informed the public of sharp increases in the prices of tortillas (100 percent), gasoline (60 percent), and electricity (30 percent). Still, the news emanated from the Ministry of Commerce, not the presidency, thus diffusing its impact. "The announcement read," in the words of a U.S. official, "like an anonymous comment delivered to nobody on a brown paper bag. It lacked any official character, certainly any connection with López Portillo. It signaled to those who knew, that inflation would run unchecked."[49] On August 5, 1982, the Finance Ministry promulgated a two-tier exchange system: a "preferential" rate of 50 pesos to the dollar reserved for crucial imports, interest and principal payments on public and private foreign debt, and other priority transactions; and an "ordinary" rate of 70 pesos to the dollar for exports, tourists entering the country, and payments of interest and principal on foreign currency bank deposits in Mexico.[50]

Rescue Effort and Austerity

The unwillingness of López Portillo's regime to tighten the nation's belt generated large government deficits, an unrealistically high peso in relation to the dollar, and prodigious capital flight by elements of the upper and middle sectors who were disturbed by the president's unpredictable behavior. Meanwhile, private and public debt service was devouring 64 percent of the country's export earnings. Only a rescue scheme, fashioned by the Reagan administration in response to a desperate August 1982 visit to Washington by Finance Secretary Jesús Silva Herzog, a liberal-rationalist, prevented Mexico's defaulting on its external debt. The bailout included a $1.7 billion guarantee from the U.S. Commodity Credit Corporation to finance grain exports to Mexico by North American farmers; a $1 billion prepaid oil sale by Pemex on remarkably attractive terms to the U.S. Strategic Petroleum Reserve whose purchases of Mexican crude would increase from 50,000 bpd to 190,000 bpd by mid-1983;[51] $1.85 billion in emergency credits negotiated in Basel through the Bank of International Settlements, which acts as the central bankers' central bank; and a fresh line of credit from private banks, which agreed to a ninety-day moratorium on principal repayments of approximately $10 billion, provided Mexico reached an agreement with the IMF.

While negotiations with the IMF were hanging fire for approximately $5 billion, López Portillo displayed his revolutionary credentials by nationalizing the private banking system and renaming Tello as head of the Central Bank of Mexico. He justified the September 1 nationalization, which caught all but the chief executive's closest advisors by surprise, on the grounds that the banking community had "betrayed" Mexico by facilitating speculation against the peso, which had lost 75 percent of its value during 1982. "I can affirm" he said, "that in recent years a group of Mexicans, led, counseled and supported by private banks, have taken more money out of the country than all the empires that have exploited us since the beginning of our history."[52] But in fact it was gross financial mismanagement, not a plot by banking gnomes, that had sparked the capital flight. This "profoundly revolutionary measure" by an outgoing leader elicited memories of Echeverría's land seizures. It appeared designed to arrest the plummeting popularity of the chief executive, identify a scapegoat for his earlier failure to remedy the country's ills, secure for himself a place alongside Cárdenas in his country's pantheon of heroes, and mollify the left. At the behest of the PRI, some 300,000 workers, peasants, and civil servants flocked to Mexico City's central plaza to praise the intrepid action of "the patriotic president." In the opinion of one writer, "López Portillo was looking increasingly like a bullfighter awarded both ears and the tail" because of his undaunted move.[53] Demagoguery reached its zenith when the president began collecting "voluntary" contributions from laborers, campesinos, and government employees to compensate the owners of Mexico's banks—a program in which the military refused to participate.[54]

A common thread joined the nationalization speech to earlier discourses by López Portillo (August 5, 1982, on the dual exchange rate) and Silva Herzog (August 17, 1982, on the financial crisis). They urged patriotic unity and individual sacrifice amid critical conditions, while largely shifting the blame for the economic malaise to forces "beyond our control." An analysis of these speeches finds the blame for these uncontrollable forces apportioned between specific groups and world conditions. The former included bankers, multinational corporations, bad "denationalized" Mexicans, orthodox technocrats, and pressure groups; the latter embraced the oil glut, the fall in petroleum prices, the decline in the export price of other Mexican primary products, the reduced quantity of petrodollars available to major world banks, attractive interest rates in the United States that drew Mexican capital north of the border, and higher interest rates and more rigorous terms on loans made abroad.[55] As López Portillo sanctimoniously expressed it: "We haven't sinned, either as a government or as a country, and we have no reason to engage in acts of contrition."[56]

A New Team Takes Over

The oil glut, fall in petroleum prices, and brush with bankruptcy produced a change in Mexico's conception of its international role. As indicated in figure 2, a lag occurred between the appearance of a buyer's market for oil and Mexico's defining itself as a Responsible Debtor. After all, several months elapsed before officials in Mexico City realized that the oil surfeit was not a transient phenomenon, but one that might bedevil the world market for years to come. Nevertheless, during the transition López Portillo, a lame duck who faced ever sharper domestic opposition to his unabashed support of "subversives," began to temper his policy. For example, in late 1981 he refused Mitterand's request to back publicly France's announced sale of $17 million in arms to the Sandinistas. Shortly thereafter, the Mexican leader declined a European democracy's private invitation to collaborate in sending funds to the FMLN-FDR in El Salvador.[57] López Portillo's gradual move away from defending revolutions was complemented by his trying to play the role of Honest Broker between the United States and the region's leftist movements. A case in point was Mexico's joining Venezuela in September 1982 to send letters to the presidents of Nicaragua, Honduras, and the United States urging them to negotiate their differences. Mexico even admitted that Nicaragua was not entirely without fault in the regional crisis, particularly with respect to tensions with Honduras. In his final address to the Mexican Congress, López Portillo talked about the "radical importance" of one development: "for the first time in history, Mexico has been converted into a sought out and accepted interlocutor by the United States in matters not directly bilateral."[58] Reality contradicted these glowing words, for the problem with his country's playing a mediating function became evident in February 1982 when Secretary of State Alexander Haig turned thumbs down on López Portillo's offer, made in Managua, to mediate the U.S.–Sandinista conflict in Nicaragua, U.S.–Cuban antagonism, and the Salvadoran civil war.

De la Madrid and Pragmatism

Fortuitously, the new energy conditions coincided with de la Madrid's ascent to the presidency, bringing to power a cautious technocrat who, while a defender of a mixed economy, believed that the state had become too interventionist and that greater reliance on market forces was needed. In addition, de la Madrid preferred accommodation to confrontation in international affairs. Having earned a master's degree in public administration at Harvard University, the forty-seven-year-old leader knew more about the United States than any of his predecessors. He appeared to comprehend that Washington, perceiving an economically healthy and politically stable Mexico to be vital to its own interests, had extended a helping hand rather than a menacing fist during the 1981–1982 financial crisis.

In order to promulgate more pragmatic policies, the new chief executive brought with him his own *camarilla,* an informal—yet highly personal—entourage based on reciprocal commitments to advance the careers and enhance the image of its members, who first worked together in the Bank of Mexico. Comprising this team were Finance Secretary Silva Herzog, Pemex Director-General Mario Ramón Beteta, Central Bank President Carlos Mancera Aguayo, and Nacional Financiera President Gustavo Petricioli Iturbide. Table 4 outlines their relationship, which—until 1977—found Beteta, the oldest of the group, holding senior posts.

The appearance at center stage of de la Madrid and his team dominated by liberal-rationalists displaced most nationalist-populists from policymaking positions. Tello devoted himself to writing and teaching before being named envoy to Portugal in 1987, and Oteyza watched the charting of Mexico's new course from Ottawa, where he served his country as ambassador. While less outspoken than Tello and Oteyza, two men in de la Madrid's nineteen-member cabinet expressed a nationalist-populist viewpoint. These were Francisco Labastida Ochoa, whose Ministry of Energy and Parastatal Industries (SEMIP) supervised both energy prices and hundreds of government-run corporations, and Bernardo Sepúlveda Amor, who succeeded Castañeda as foreign secretary and is renowned for his nationalist-populist ideals, anti-U.S. sentiment, and sympathy for an assertive Mexican foreign policy keyed on associating Mexico's interests with those of Third World nations. Under de la Madrid, the Foreign Ministry continued to take exception to U.S. foreign policy, although less stridently than during the Echeverría and López Portillo administrations. Criticism focused on the armed intervention in Grenada, the CIA-backed mining of Nicaraguan ports, military maneuvers in Honduras, and inadequate support for the Contadora process. Even more conspicuous was the SRE's role in U.N. votes, which will be analyzed in chapter 4. Nonetheless, the importance of the Foreign Ministry in decision-making councils waned as the growing salience of economic issues enhanced the influence of the Finance and Commerce Ministries, the Central Bank of Mexico, Pemex, and other bastions of liberal-rationalist thought. The fact that the politically ambitious Sepúlveda's name failed to appear on the first public short list for the PRI's presidential nomination offered compelling evidence of the marginality of his SRE in policymaking.[59]

Upon assuming office, de la Madrid set forth a ten-point "economic reorganization program" in which he underscored the need for austerity and criticized his country's profligacy—a refreshing approach in a nation whose previous leaders frequently blamed the "colossus of the north" for real and perceived injuries. "All Mexicans must contribute to [resolving the crisis], aware that there is no foreign aid or magic solution to replace that effort." In an unprecedented reproval of his predecessor, de la Madrid rejected "financial populism," while endorsing "structural changes" in Mexico's mixed economy to "combat failures, insufficient domestic savings, low productivity, insuffi-

TABLE 4
Positions Held by Members of the Central Bank of Mexico Camarilla

	Mario Ramón Beteta Monsalve	Miguel de la Madrid Hurtado	Jesús Silva Herzog Flores	Gustavo Petricioli Iturbide
Date and place of birth	July 7, 1925, Federal District	December 12, 1934, Colima	May 8, 1935, Federal District	August 19, 1928, Federal District
Education	Law (UNAM, 1948); Economics (Wisconsin, 1950)	Law (UNAM, 1957); Public Administration (M.P.A., Harvard, 1965)	Economics (UNAM, 1959; M.A., Yale)	Economics (Technical Institute of Mexico 1952; M.A., Yale, 1958)
1958–1964	Subgerente and gerente, Bank of Mexico[1]	Advisor to administrator, Central Bank of Mexico, 1960–1964	Department of Economic Studies Central Bank of Mexico, 1956–1960; Inter-American Development Bank, 1962–1963	Economist, Central Bank of Mexico 1958; director of Technical Office, Central Bank of Mexico[2]
1964–1970	Director-general of credit, Finance Ministry	Deputy director-general of credit, Finance Ministry	Director of Technical Office, Central Bank of Mexico, 1964–1968; coordinator, Central Bank of Mexico, 1969–1970	Gerente, Central Bank of Mexico; director-general of treasury studies, Finance Ministry, 1967–1970
1970–1976	Subsecretary of credit, Finance Ministry, 1970–1975; secretary, Treasury, 1975–1976	Subdirector of finances, Pemex, December 1970–April 1972; director-general of credit, Finance Ministry, May 1972–1975; Subsecretary of credit, Finance Ministry, 1975–1976	Director-general of credit, Finance Ministry, 1970–1972; director-general of the National Institute of Housing, April 1972–1976	Subsecretary, Finance Ministry, 1970–December 1974; subdirector, Central Bank of Mexico, 1975–1976
1976–1982	President, Banco Mexicano Somex	Subsecretary of credit, Finance Ministry, 1978–1979; secretary, Planning and Budget, 1979–1982	Central Bank of Mexico, 1977–1978; director-general of credit, Finance Ministry 1978–1979; subsecretary of credit, May 1979–March 1982; secretary, Finance Ministry, March–December 1982	President, National Securities Commission
1982–1988	Director-general, Pemex, 1982–February 1987	President of the Republic	Secretary, Finance Ministry, 1982–August 1986	President, Nacional Financiera, S.A., 1982–August 1986; secretary, Finance Ministry

Sources: Roderick A. Camp, *Mexican Political Biographies, 1935–1981,* 2d ed. (Tucson: University of Arizona Press, 1982), pp. 32, 81, 127, 188; Mario Ramón Beteta, curriculum vitae, distributed by Pemex, October 1981.
1. Beteta entered the Central Bank of Mexico as an economist in 1949.
2. Petricioli entered the Central Bank of Mexico as an economist, Department of Economic Studies, in 1948.

cient competitiveness of our products abroad and social inequality." While committed to an activist state, the new chief executive stressed the need to curb the growth of public spending, to respect the prerogatives of "responsible, nationalist businessmen," and to avoid "squandering, waste, and corruption." "Public offices," he averred, "should be no one's booty."[60]

De la Madrid's commitment to such structural reforms as trimming subsidies, eliminating inefficient government-owned firms, pursuing realistic prices, and decreasing the state's economic status was crystallized in his support of a November 1982 agreement with the IMF. Under this accord, the Fund would provide Mexico with a $3.84 billion extended fund facility credit, as well as $40 million to $880 million from its compensatory financing facility, established to assist countries suffering from declining exports. Upon the request of the IMF's executive director and the U.S. secretary of the treasury, 530 private banks supplied Mexico with $5 billion in credit to carry the country through 1983. In addition, major foreign banks provided a bridging loan of $33 million to meet debt obligations falling due at the end of February 1983, until the first payment of $1.7 billion from the large credit was available. In return for the infusion of credits, Mexico agreed to slash its budget deficit from 16.5 percent of GDP in 1982 to 8.5 percent in 1983, with further cuts contemplated for 1984 (5.5 percent) and 1985 (3.5 percent). Moreover, it would limit foreign borrowing to no more than $5 billion in 1983 and take steps to build up its hard currency reserves to $2 billion. The Mexicans also pledged to modify the foreign exchange controls imposed in August 1983 with a view to moving eventually to a unified exchange rate. Further, they promised to restrain public sector wages, boost productivity, eliminate or reduce subsidies to the private and public sectors, and reduce investment in petroleum production and industrial plants. Revenue expansion through fiscal reforms was also contemplated in order to widen the tax base in a country where the poorest 20 percent of the population received 2.9 percent of total income, while the top 10 percent enjoyed 40.6 percent.[61]

Even though exaggerated by Mexican and non-Mexican officials alike, the commendable sacrifices Mexico made to achieve the goals negotiated with the IMF elicited plaudits from an international financial community eager to identify a model debtor within the Third World. The community wanted to send a signal: its readiness to restructure Mexico's loans was meant to show developing countries contemplating default or membership in a debtor's cartel that relief awaited responsible nations courageous enough to put their economic houses in order. In October 1983, *Euromoney* magazine named the dashing, motorbike-riding Silva Herzog as "finance minister of the year" for his adroit handling of the debt crisis; the *Wall Street Journal* reported that "bankers smile at the mere mention of Mexico's charismatic finance minister;" and a headline in the *Economist* adverted to "Mexico's Silva lining."[62] Similarly, Adrian Hamilton, writing in the *Observer* magazine, asserted: "Rescheduling has been expensive, but Mexico has taken its medicine with determination and effectiveness." Even more sanguine was political scientist Wayne A. Cornelius, who believed that for the rest of de la Madrid's tenure "the government . . . [would] steer a moderate course toward recovery, avoiding the extremes of further deflation and IMF-style austerity budgets, on one hand, and a return to the hyperexpansionary policies of the past, on the other."[63]

Indeed, hard times forced many Mexicans to adjust their consumption habits. Lard replaced cooking oil in the kitchens of many lower-class families who sent their youngsters to sell lottery tickets or hawk plastic brushes in Mexico City's traffic jams; members of the middle class moved to smaller apartments, reduced their use of domestic help, and enrolled their children in Mexican rather than foreign universities; and Fidel Velázquez imposed discipline on union leaders to minimize the number of strikes and antigovernment demonstrations. Unfortunately, the country's improved statistical picture in terms of a trade surplus and accumulation of foreign exchange reserves resulted from curbing imports rather than spurring production or expanding exports.

Just as it appeared that the patient was beginning to recover, Mexico began to spend feverishly in late 1984. Mexican officials and sympathetic commentators attributed the policy shift to a combination of egregious misfortune and poor economic management, such as imperfect budget controls at state enterprises. Specifically, oil prices began to slide again in the spring of 1985, depriving the government of critically needed resources. Before the government had adjusted to diminished income, the worst natural disaster in modern history—the El Grande earthquakes of September 19–20, 1985—struck Mexico City, killing upwards of 20,000 people, leaving 90,000 families homeless, and causing an estimated $3.5 billion in property damage. The economic situation was aggravated because government decision makers claimed not to comprehend the speed of expansion until the overstimulated economy had spun out of control. The boomlet produced growth of almost 7 percent, if calculated on an annual basis, in late 1984 and early 1985.

While acknowledging bad luck, many nongovernment observers explained the reflation in less sympathetic terms. They contended that de la Madrid lacked the political conviction to stay the course of economic retrenchment. Succinctly stated, he became alarmed by the victories registered in 1983 by the center-right National Action party (PAN) in major municipal contests in Durango, Chihuahua, and Ciudad Juárez, and as a result accelerated government public outlays to strengthen the PRI's standing in the mid-1985 legislative and gubernatorial elections.

While politics was an important consideration, it appeared that the government underestimated the belated, but significant, private sector expenditures when devising its own spending strategy. The spending binge was enough to collapse the stabilization program. However, it was neither wildly excessive nor accompanied by a public intracabinet donnybrook, although severe tensions characterized relations between Silva Herzog and Planning and Budget Secretary Carlos de Salinas Gotari. As a writer for the *Wall Street Journal* observed: "The public wasn't treated to the usual bloody spectacle of cabinet ministers quarreling over vastly different views of government. The current crisis is the cumulative effect of small miscues, behind-the-scenes rivalries, and a gradual erosion of faith in the regime's economic direction."[64]

In part de la Madrid may have boosted government spending to vouchsafe the PRI's success in the July 7, 1985, gubernatorial and legislative elections. Nevertheless, the official party that manipulated elections during periods of prosperity was loath to relinquish control over key offices amid an economic recession. This attitude was particularly evident in the races for the statehouses of Sonora and Nuevo Leon where the PAN fielded attractive, energetic candidates. As a result, officials engaged in blatant electoral chicanery to ensure the victory of PRI nominees. Consequently, Mexico emerged from the balloting in the worst possible position: it had jettisoned a painful though effective austerity plan while reinforcing its image as a nation suffused by fraud and corruption.

The reflation of late 1984 notwithstanding, de la Madrid followed a pragmatic line on international economic questions. In mid-1985, Mexico reversed a previous decision and granted permission to International Business Machines to build a wholly foreign-owned microcomputer plant in Jalisco state near Guadalajara. Such a move, though highly beneficial to Mexico because of the attractive concessions exacted from the U.S. firm, was interpreted as an olive branch extended to foreign investors, who, understandably, had shown little desire to risk capital in the Mexican economy. In November 1985 de la Madrid further demonstrated his commitment to "structural reform" by announcing his intention to lead Mexico into GATT, a ninety-one-member, Geneva-based organization designed to stimulate world commerce by reducing tariffs, quotas, import permits, and other barriers to trade. López Portillo took his country to the brink of entry in 1980, only to withdraw in the face of shrill opposition from a curious coalition of nationalist intellectuals and small- and medium-sized elements of the business community. De la Madrid's initiative was designed to liberalize his nation's heavily protected, oil-dependent, statist economy. Membership in GATT, formalized in August 1987, provided an important assurance to the private sector and the international financial community of an attempt to promote continuity between de la Madrid's administration and that of his successor, who would take office on December 1, 1988.

The 1981 and 1982 Pemex sales to the U.S. Strategic Petroleum Reserve, López Portillo's attempts to serve as an Honest Broker in Central America, the adoption of an IMF stabilization plan, and the imposition of austerity measures all reflected Mexico's shift from Regional Leader to Responsible Debtor in its external affairs. The transition—consummated in the mid-1980s when de la Madrid dispatched an ambassador to San Salvador, cut back economic aid and oil shipments to the Sandinistas, led his country into the GATT, expelled several Cuban diplomats, and refrained from visiting Managua or Havana— did not prevent Mexico's assiduously seeking to influence the course of Central American affairs. As I have mentioned before, in the early 1980s Mexico largely abandoned the unilateral demarches that often produced clashes with the United States in favor of bilateral or multilateral ventures. The speeches of SRE spokesmen in international bodies aside, Mexico generally attempted to work with other nations in a less confrontational manner than before. This brokership approach characterized Mexico's participation in the Contadora Group, formed in January 1983 when Sepúlveda met with his Venezuelan, Colombian, and Panamanian counterparts to seek a diplomatic alternative to the rapidly escalating armed strife in the region. Although various schemes to diminish arms stockpiles, curb the size of military forces, secure borders, and eliminate military advisors have won approval of one or more key adversaries in the conflict, the Contadora process has not succeeded partly because Nicaragua resists on-site verification and demilitarization and partly because the Reagan administration has committed itself to the contras in an effort to enfeeble the Sandinista government and enhance Washington's influence in the region.[65] While no one had pronounced the process dead, an astute scholar noted in late 1986 that "Contradora is clearly comatose and its vital signs are ebbing."[66]

Even while championing the multilateral Contadora initiative, de la Madrid pursued the role of Honest Broker in urging a direct U.S.-Nicaraguan dialogue. This effort bore fruit when, following the Mexican president's mid-May 1984 meeting with Reagan, Secretary of State George P. Shultz made a brief stopover at the Managua airport for the first of nine high-level talks between the United States and Nicaragua. That most of the sessions took place in the Mexican resort of Manzanillo testified to the host country's importance in bringing about the exchanges.[67] However, the sessions did not produce an understanding between the two nations.

On foreign debt matters, Mexico shunned action that would endanger its relationship with industrialized countries and international banks. While join- ing the Cartagena Group—a loose, eleven-member association whose collec- tive obligations accounted for more than 80 percent of the $360 billion owed by Latin American states—Mexico refused to support either debt repudiation, as championed by Castro, or limiting debt servicing to a percentage of export revenues, as advocated by Peru's President Alan García. Nevertheless, Mexico greeted with skepticism the so-called Baker Plan, enunciated in October 1985 by U.S. Treasury Secretary James A. Baker III to emphasize economic growth in Third World economies rather than retrenchment. The plan called on commercial banks to expand significantly lending to poor countries, while the World Bank as a promoter of development would replace the IMF as the key international agency dealing with the debtor states. In exchange for an infusion of fresh resources, these nations would be expected to close inefficient public firms, to tumble tariff walls, to diminish subsidies, to revise tax structures, and to encourage foreign investment. Mexican misgivings focused on funding levels that were deemed inadequate, as well as the proposal's failure to address high interest rates, the protectionism practiced by developed nations, flaccid commodity prices, and other problems facing most Third World countries. Moreover, as political scientist Jorge G. Castañeda, son of López Portillo's foreign secretary, argued: "But new funds, even if they were available in sufficient quantity—a doubtful proposition—would only postpone the problem, compounding it: New debts this year mean more interest to pay next year and every year after. These are the problems that the Baker plan simply does not address."[68]

Such reservations permeated a December 1985 emergency session of the Cartagena Group held in Uruguay, which had been called by Mexico and Venezuela in reaction to plunging oil incomes. Officials from these two countries joined others on the organization's steering committee (Colombia, Brazil, and Argentina) in a communiqué, declaring that in the case of individual debtor states "substantial modifications to existing debt agreements could no longer be postponed, in particular with regard to current interest rate levels."[69] Nevertheless, a careful reading of the Declaration of Montevideo reveals that the proponents of conciliation triumphed over enthusiasts for confrontation. Absent from the final version were such fiercely debated points as demands for a 3 percent reduction in the region's interest rates and a limitation on profit margins on new loans to just one-half of one percent. Another example of moderation was the refusal to transfer the secretariat of the Cartegena Group to Peru, whose president is a firebrand on debt questions.

Argentina was especially disappointed at Mexico's unwillingness to lead the charge for such assertive goals as the unilateral reduction of annual interest payments. Yet, while desiring hemispheric company in financial misery, Mexican leaders understood that their nation's economic and strategic importance augured well for preferential treatment by Washington and the international financial community—advantages that radicalism and bravado would obviate.

Such treatment appeared in the form of a July 22, 1986, pact that Finance

Secretary Petricioli, who had replaced Silva Herzog six weeks before, signed with the IMF on "national territory," the Mexican embassy in Washington. The eighteen-month "sovereign program" not only provided for $1.7 billion in immediate Fund credits, but it also opened the way to an additional $10.3 billion from the IMF, World Bank, Western governments, and commercial banks. The projected $6 billion of additional commercial bank credits was contingent upon Mexico's reaching agreement with the IMF. The giant money center banks also insisted that their disbursement of fresh credits to Mexico await the participation in the agreement of some 200 regional banks, some of which were reluctant to deepen their involvement in Mexico's floundering economy. New loans would be paid over twelve years, with a seven-year grace period, and an interest rate of 13/16 of a percentage rate above LIBOR. Payments on $43.7 billion of existing debt were stretched over a twenty-year term, with a seven-year grace period at the new LIBOR rate. In a major breakthrough, explicit growth rates and long-term structural adjustments eclipsed the traditional objectives for tighter monetary and fiscal policies. For instance, the World Bank agreed to become more active in financing the expensive process of structural change and to assist in the further liberalization of imports and the reduction of the size of the public sector—the benefits of which would not accrue for several years.

Especially important to Mexico were contingency mechanisms with respect to oil prices and investment.[70] The plan assumed that the average oil price would hover between $9 and $14 per barrel in the September 1986–March 1988 period. Should prices drop below $9 per barrel, Mexico would be allowed to solicit additional financing to compensate for 100 percent of the shortfall caused by each dollar decline in oil prices for a nine-month period. After nine months, and to the end of the program, the nation's borrowing compared to the size of the shortfall would progressively diminish as economic adjustments gradually offset a portion of the lost petroleum income. Overall potential funding would be $1.8 billion, of which $1.2 billion would come from commercial banks (Investment Support Facility) and $600 million from the IMF (Oil Contingency Mechanism). Should average oil prices exceed $14 per barrel, Mexico's external financing requirements would be reduced in proportion to the additional revenue.

Mexico's failure to recommence growth in the first quarter of 1987 would activate a Growth Contingency Financing Facility. Through this mechanism, the nation could borrow as much as $500 million for investment projects previously identified and appraised by the World Bank. Commercial banks would provide these monies, half of which would be guaranteed by the World Bank.[71]

Apparently, only a threat to halt disbursement of interest payments to creditors broke a protracted stalemate between Mexico and the so-called theologians in the IMF.[72] Yet, the agreement with the financial community was the culmination of an education process for bankers, international economic officials, and developing states, demonstrating that world leaders were sufficiently flexible to learn from experience when required. One analyst's observation about the scheme was especially pertinent to Mexico's emergence as a Responsible Debtor: "What started as a daredevil game of poker—with the fates of governments and major banks to play for—has turned into a tense, but rational, chess match."[73] Performing as a Responsible Debtor provided de la Madrid sufficient financing to satisfy basic economic needs through the end of his term. Doubtless, Salinas will have to embark upon yet another round of debt negotiations, for negative growth in 1986 combined with slow growth and rampaging inflation in 1987 darkened Mexico's economic prospects.[74] Still, the eighteen-month program bought precious time for bankers and the Mexican government. The former could continue to diminish their exposure in Third World countries; the latter could make at least marginal headway in launching a newly unveiled Program for Recovery and Growth while reforming its economy. Mexico's accomplishing the structural adjustments contemplated by the IMF agreement depended on an efficient Pemex's earning the necessary foreign exchange to finance these changes.

Mexico's Oil Industry Under
de la Madrid: A New Pemex?

Herculean problems beset Mexico's oil industry when de la Madrid became president in late 1982. Díaz Serrano, director-general of Pemex during most of López Portillo's administration, had encouraged wildcat drilling and frenetic development of newly discovered onshore and offshore reservoirs to make Mexico the world's fourth largest oil producer. He also catalyzed the construction of new refineries, pipeline systems, processing facilities, deep-water ports, and the largest petrochemical complex in Latin America. The emphasis was on results, not means, and the motto for Díaz Serrano's buccaneering style seemed to be: "Waste anything but time"—for a production élan took precedence over careful, professional management of the oil sector, which became a beehive of activity. Consequently, enormous waste sprang from inadequate or haphazard procedures in planning, budgeting, accounting, purchasing, personnel management, inventory control, maintenance, and environmental protection.

There is a popular saying in Mexico that "Oil, a gift of the gods, is the temptation of the devil." The freewheeling atmosphere of the late 1970s nurtured the corruption that had long suffused the petroleum industry, but was exacerbated by the enormous projects and huge earnings arising from the petroleum boom. Leaders of the 176,734-member Oil Workers' Union (SRTPRM), Latin America's economically most powerful labor organization, demonstrated an unusual talent for enriching themselves and their organization from the national patrimony. Their wrongdoing included job-selling, property theft, collecting the salaries of nonexistent employees, coercive thuggery against dissident members and other union foes, and exacting large commissions for work contracted from Pemex that was never completed, done poorly, or subcontracted to third parties.[1] This peculation reached into Pemex's highest echelons, perhaps into the presidential palace itself. The popular outrage over these highly publicized abuses may have persuaded López Portillo to spend the early part of his successor's *sexenio* in Europe.

Also disturbing to the new administration was the legacy of fragmented decision making characterizing energy policy in general and foreign sales in particular. Although the chief executive could make the final determination of prices and customers, López Portillo largely delegated that authority to Díaz Serrano, who was forced to resign following the June 1981 reduction of crude prices. The Ministry of Patrimony, under whose formal jurisdiction Pemex fell, played a major role in Mexican-OPEC relations; the Ministry of Foreign Relations shaped policy with respect to concessionary oil sales to Nicaragua and nine other beneficiaries of the San José Accord; and the Ministry of Finance and Public Credit, responsible for levying and collecting taxes, exerted a strong influence on domestic energy charges.

In view of these conditions, de la Madrid said in his inaugural speech:

Energy resources are an ineluctable part of our nation's patrimony. We still exploit oil in order to continue along the road of development, taking care to use it for the benefit of present and future generations, without considering petroleum a panacea that diminishes efforts in other aspects of our development. We must pay greater attention to adequate planning and the honest and efficient development of our petroleum industry, the fundamental pillar of our economic nationalism.[2]

This chapter describes the strategy that the de la Madrid administration used to fashion a "new Pemex," assesses the domestic impact of this policy, especially regarding management goals, discusses the moral renovation program for combating corruption, and analyzes Mexico's international policy, with emphasis on the making and implementation of decisions.

Strategy for a "New Pemex"

To accomplish changes in Pemex, de la Madrid named as director-general Mario Ramón Beteta, a wealthy and aristocratic fifty-five-year-old lawyer-economist, who was the son of a revolutionary general who served under President Cárdenas. After earning a law degree at UNAM and a master's degree in economics at the University of Wisconsin, Beteta spent more than twenty-five years in the financial bureaucracy. His uncle was a finance secretary and professor who strongly influenced the generation of technocrats who came to dominate the PRI in the early 1980s. Himself a quintessential technocrat, Beteta won distinction in a series of key posts in the Bank of Mexico and the Finance Ministry, which he headed from September 1975 to December 1976.

The first devaluation of the peso in more than two decades demanded a change in secretaries. López Portillo offered Beteta either of two new posts, the ambassadorship in Washington or the presidency of the Banco Mexicano Somex, a state industrial credit institution. The outgoing finance secretary accepted the latter, even though it was a decided step downward in a career previously characterized by continuous upward mobility. He made this move because service abroad could be interpreted as acceptance of the blame for a devaluation that was inevitable. He believed it imperative to remain at home to defend his reputation.

In Beteta, the president found a man reputed for his honesty, competence, managerial skill, experience, and sound, conservative economic philosophy. Equally important, the new director general was not only a fellow technocrat, but also a confidant on whose loyalty de la Madrid could count. As discussed in chapter 2, the two men belonged to the same *camarilla,* and their careers had been remarkably similar.

Beteta readily accepted the challenge to improve the performance of the world's twelfth largest corporation, as indicated in table 5, that in 1982 generated 7 percent of Mexico's GDP and 30 percent of the federal government's income. In his first annual report as director-general, delivered in de la Madrid's presence on March 18, 1983, he underscored the "disparity" in the growth of Pemex compared to its ability to manage, plan, and process information—a divergence aggravated by economic conditions at home and abroad and by "human failure within the institution often implying corrupt practices."[3]

Pemex's leader pledged to promulgate the president's highly publicized "moral renovation" in order to fight wrongdoing in the monopoly whose reputation had been "sullied" by the "lamentable conduct of . . . [a] few disloyal" employees within an excessively large bureaucracy. "Our society's moral renewal concerns us directly," he said, "because our petroleum industry

TABLE 5
Pemex's Ranking Among International Corporations, 1962–1985

	Pemex's Assets (in dollars)	Ranking		Pemex's Assets (in dollars)	Ranking
1962	976,527	50	1974	3,738,416	72
1963	1,022,451	58	1975	5,563,946	62
1964	1,132,640	62	1976	3,069,423	70
1965	1,169,881	64	1977	8,593,976	73
1966	1,289,716	66	1978	n.a.	66
1967	1,578,711	67	1979	n.a.	39
1968	1,829,334	65	1980	31,804,157	20
1969	2,036,355	66	1981	45,072,720	11
1970	2,162,575	71	1982	20,712,500	21
1971	2,230,445	69	1983	32,598,710	12
1972	2,439,736	76	1984	37,432,599	9
1973	2,960,848	90	1985	36,633,436	9

Source: Fortune, 1963–1985.

can no longer tolerate the obscuring of its values, nor this expression of vices of a few negative elements, not its own."[4]

Among the tasks facing Petróleos Mexicanos, Beteta cited (1) increasing the motivation to produce, (2) reassigning personnel to assure greater efficiency, (3) decentralizing the firm's activities, (4) avoiding "at all costs" unnecessary hiring, and (5) emphasizing training and development programs whose standards had declined in recent years even as the number of activities grew.[5]

Parallel to the director-general's stated goals, the president approved on August 13, 1984, a National Energy Program (NEP) for 1984 to 1988, the eight goals of which furnished a blueprint for Mexico's energy future:[6]

1. Achieving long-run self-sufficiency, which is the highest priority of the sector, in a manner consistent with satisfying the energy needs of the country at the lowest possible cost. This goal anticipates diversifying the nation's energy sources in order to fortify national sovereignty.
2. Contributing to national development by generating foreign exchange, taxes, and economic activity. Inasmuch as the strategically situated energy sector is the largest consumer of capital goods, industrial inputs, and services, its greater and more efficient integration with the economy will advance industrial development and promote modernization.
3. Assisting in social development by expanding protection and avoiding regional and environmental disequilibrium. The sector will reinforce its contribution to social development by supporting priority activities— specifically, extending to the growing number of dispersed rural and urban poor the benefits associated with available energy supplies and promoting balanced regional development and environmental protection in order to construct an egalitarian society.
4. Conserving energy and encouraging its efficient use in production, distribution, and final utilization, while diminishing neither productivity nor the quality of life.
5. Attaining a more rational energy balance in a country 92 percent dependent on hydrocarbons by developing hydropower, coal, uranium, and geothermal energy.
6. Improving Mexico's scientific and technological capability to enhance national independence and promote a healthy economy and an egalitarian society.
7. Achieving a more efficient and better integrated energy sector. This goal involves increasing the productivity and administrative, technical and operating efficiency of Pemex, the Federal Electricity Commission, and other energy suppliers congruent with the principles of austerity and budgetary discipline imposed by the federal government.

8. Helping to strengthen the world oil market in accordance with national interests, which entails defending a fair price for the country's natural resources through mechanisms of international coordination and cooperation.

Beteta brought in his own management team to forge the "new Pemex"— namely, to attain the objectives set forth in his report, in the National Energy Program, and in rolling five-year plans that Pemex revises and brings up to date annually. Of the seven subdirectors whom he designated, only two had previous experience in the monopoly, while four followed Beteta from the Banco Mexicano Somex.[7] In addition, the director-general recruited several thousand men and women unencumbered by previous service in the state firm to assume scores of second- and third-tier assignments. They were even more likely than the subdirectors to boast foreign educations and backgrounds in economics, finance, accounting, or management rather than in petroleum engineering, chemical engineering, or other energy-related professions. It was argued that their training would facilitate the auditing of accounts and the monitoring of transactions. After all, the monopoly's old financial statements appeared designed to confuse rather than enlighten so that its reported earnings—$16 million after taxes in 1981 on revenues of $19.2 billion—were deemed "meaningless."[8] Details on costs and spending were shrouded in mystery. Many of the recruits entered the proliferating "coordination" offices established between the seven subdirectorates (*subdirecciones*), responsible directly to the director-general, and the managerial level (*gerencias*). The experience of the new employees and the creation of so many coordinating offices reflected Beteta's conviction that the monopoly was underadministered compared to multinational oil companies.[9] As it turned out, the influx of technocrats provided ammunition to Beteta's critics who condemned the growth in administrative personnel when the need for belt-tightening appeared imperative.

Domestic Impact

It will take years to evaluate Beteta's success in achieving the objectives of the 1984 NEP such as long-run self-sufficiency, diversification and improved integration of the energy sector, pollution control, and higher living standards for Mexico's 84 million citizens.

In the face of retrenchment imposed by financial exigency, the monopoly stacked drilling rigs, laid aside blueprints to expand refineries and pipelines, and delayed construction of such projects as the mammoth La Cangrejera petrochemical complex in Veracruz state. The austerity strengthened Beteta's support for higher domestic prices to rationalize consumption, promote conservation, and generate critically needed funds for both the government and the oil sector. Yearly price changes soon gave way to periodic adjustments, which by 1986 led to monthly increases linked to both the evolution of the economy and the international market. At the beginning of 1986, the government instituted such increases for diesel fuel, No. 6 fuel oil, liquified gas, and natural gas. In August 1986 it began monthly increases for regular (Nova) and high-test (Extra) gasoline. The results were prices that approached those in the United States. For instance, in the spring of 1987 Pemex sold regular gasoline for 68.1 cents per gallon and high-test for 78.9 cents per gallon, while comparable U.S. prices were $1.06 per gallon and $1.11 per gallon, respectively.[10]

Beteta claims that these domestic price increases have combined with higher export revenues and reduced operating costs to improve the monopoly's financial standing during the first half of his tenure. As a result, earnings were 49 percent higher in 1984 compared to 1983, even though the volume of sales was virtually the same in the two years. Pemex's real net savings, referred to by businessmen as net profits before taxes, amounted to 2.7 trillion pesos or 72 percent of sales. This permitted the monopoly to pay 1.7 trillion pesos in direct taxes and duties, a sum equivalent to 46 percent of its total sales. At the same time, Petróleos Mexicanos reported a 30 percent fall in average cost of producing a barrel of oil, using constant prices.[11] The picture darkened dramatically in 1986 when export revenue amounted to $6.7 billion, just 42 percent of the previous three-year annual average. By repaying almost $5 billion in foreign loans between 1983 and 1987, the monopoly lowered its external debt from almost $20 billion to $15.35 billion. Fixed assets grew to 8.5 trillion pesos in 1985; and by decreasing foreign currency liabilities while augmenting its net worth, the ratio of the corporation's own resources to those borrowed rose from 43 to 57 percent. During the 1983–1986 period, Pemex was able to finance internally 70 percent of its operations and investments, a figure that was only 20 percent during the López Portillo sexennium. However, plunging world oil prices meant that the firm financed internally only 40 percent of its total assets in 1986, thereby limiting its "capacity to absorb more debt in the future."[12]

Starved for funds, the government viewed Pemex, sometimes referred to as the "petroleum milch cow," as an ever more attractive source of revenue, and the Ministry of Planning and Budget (SPP) was the lead agency in seeking additional resources. Consequently, the monopoly, which provided some 30 percent of all federal income in 1982, accounted for 50 percent of taxes collected by the Mexican treasury in 1986—an amount equal to 63 percent of Pemex's gross revenue. This contribution took its toll on Pemex activities, the most important of which was the discovery and development of oil fields. During López Portillo's administration (through 1982), the state firm drilled an average of 78.3 exploratory wells and 180 developmental wells annually.

During the first three years of de la Madrid's term, the corresponding figures were 65.3 and 218.5, respectively. (See table 6.)

Moreover, the five-year plan anticipated completion of 400 exploratory wells; still, after three years, Pemex had drilled only 146 wildcat wells, which constituted just under half of its goal. It would be wrong to conclude from the increased success rate—32.4 percent (1986) compared to 23.7 percent (1984)—that oil had become easier to find in Mexico. To begin with, more wells had to be drilled in 1986 to discover ten new fields than were required in 1954 to find 13 new fields. Similarly, a greater number of wells were needed to locate the same volume of oil. These facts indicate that the cost of finding a barrel of oil rose because of the greater exploratory effort required.[13] Meanwhile, in 1984 the firm registered an 85 percent success rate as it drilled 228 development wells, a figure that represents approximately one-quarter of such wells anticipated during the five-year period; in 1986 a 76 percent success rate attended the drilling of 178 developmental wells.[14]

In November 1985 Abundio Juárez Mendez, the subdirector for primary production, warned that Pemex required more and better equipment in order to discover and develop oil needed to replace the 1 million bpd from existing wells that would not be available in 1990. "High exploration costs and declining crude oil prices have a negative impact on reserves," he noted. "These facts make it necessary to keep reserves at the highest level possible."[15] One year later, he warned that additional investment was required to guarantee the continuance of exploration and development programs at the appropriate level. "In the oil industry," he said, "projects have a long maturity period. Thus

TABLE 6
Pemex Well Drillings, 1977–1985

	No. of Rigs	Exploratory Wells	Success Rate[1] (%)	No. of Rigs	Develop-mental Wells	Success Rate (%)
1977	n.a.	79	38.0	n.a.	228	77
1978	n.a.	83	33.7	n.a.	223	81
1979	n.a.	83	36.1	n.a.	250	76
1980	74	85	42.0	141	349	88
1981	70	70	34.3	n.a.	342	87
1982	72	70	25.7	n.a.	288	81
1983	78	65	26.2	104	249	88
1984	85	59	23.7	110	228	85
1985	92	69	27.5	108	219	81
1986	83	68	47.0	80	178	86

Source: Petróleos Mexicanos, *Memorial de labores,* 1977–1986.
1. These figures, which exclude geologic wells *(sondeos estratigráficos)*, are inflated because of the inclusion of the extention and deepening of existing wells.

decisions made today, will have either a positive or negative effect on the volume of hydrocarbons produced over a long term."[16] Juárez may have attempted to send a signal to the country, in general, and to apply pressure on SPP, in particular, by announcing a decline in reserve levels for the first time since 1972. Figures published by the state corporation revealed that on December 31, 1984, proven reserves totaled 71.75 billion barrels of liquid hydrocarbons (oil, natural gas, condensates, etc.) compared to 72.5 billion barrels one year before; on December 31, 1985, this level fell to 70.9 billion barrels; and on December 31, 1986, it declined to 70 billion barrels.[17] Additionally, between 1986 and 1990 Pemex expects to add 5.2 billion barrels to reserves, largely from prolific offshore fields. Because this volume represents only 73 percent of extraction, reserves will decline at least 4 percent during the period.[18]

Of course, even with Juárez's admonitions, the announced reserves will remain suspect until subjected to an independent evaluation. This skepticism springs from (1) Pemex's practice of employing a 5,000-to-one ratio in converting cubic feet of gas to barrels of oil in contrast to the 6,000-to-one industrywide standard, (2) its lack of rigor in calculating secondary recovery coefficients that determine the quantity of additional oil that can be extracted from a reservoir by injecting water, gas, or steam, (3) its assumption that an undeveloped deposit exhibits geological properties similar to a contiguous producing area, and (4) its ascribing 12.2 billion barrels of oil and gas the Chicontepec field in Veracruz state even though the area, distinguished by low porosity and low permeability, would necessitate drilling more wells than the monopoly has completed in fifty years of existence.[19]

The director-general reported that reduced outlays for renting foreign vessels and warehouse space have saved substantial amounts of foreign currency. A 60 percent cutback on the former—accomplished by improving port facilities at Pajaritos where a floating storage capability was developed—has afforded yearly economies of $200 million. In addition, Pemex began taking inventory of the millions of tons of equipment, some of which were of the obsolete, "Curiosity Shop" variety, at home and abroad. This effort, which brought repatriation of more than a half-million tons of goods for use at proper work sites, rendered savings of some 95 billion pesos.[20]

Beteta also commended an import substitution program accelerated in 1983 not only to encourage the domestic production of tank-trucks, tanker cars, ships, and other equipment used by Pemex, but also to foster technological innovation, decentralized production, and exports of these goods by Mexican-based firms deemed to have the "necessary industrial capability, technology, and know-how."[21] Reportedly, the manufacture-in-Mexico program had realized modest success by early 1987. A case in point was the Abkatún project that injects 1 million bpd of water in a Campeche basin offshore field to enhance output. Mexican goods valued at 8,500 million pesos comprised two-thirds of the capital equipment used in this venture, compared to a 20 percent industrywide figure. Pemex offered financial inducements to Mexican firms engaged in import substitution. As of November 1984, thirteen firms had borrowed a total of 3,850 million pesos. The program allegedly saved $550 million in the 1983–1985 period and Beteta noted with pride that imports, which represented two-thirds of Pemex's acquisitions in 1981, constituted no more than one-fifth of its purchases by 1985.[22] Unfortunately, the state firm did not submit its figures to outside auditors. Nor had there been a cost-benefit study to assay either the quality or cost of the subsidized domestic manufactures when compared to available imports. It was widely perceived among oilmen in Mexico City that most internally produced items were more expensive and of lower quality than imported goods. Of course, Beteta regarded the program as a means to induce international corporations to establish plants in Mexico not only to serve Pemex but also to export to the world market. This hope remained a pipe dream in 1988 because of the depressed condition of the oil-equipment and service industries in the United States.

The director-general's annual reports lauded progress made in fighting pollution, one of the severest problems afflicting both industrialized, traffic-clogged urban centers and rural areas where drilling operations take place. Also, contamination had damaged Mexico's inland and coastal waters; for example, unprofessional practices allowed a Pemex offshore project to produce the worst oil spill in the industry's history. Before the Ixtoc-1 well—which suffered a blowout on June 3, 1979—was capped nine months later, 3.1 million barrels of oil and 3 billion cubic feet of gas spewed into the Gulf of Campeche.[23]

Increasing sensitivity to ambient conditions appeared in the five-year plan, as well as in Beteta's assurance that his agency had signed agreements on environmental protection with the Ministry of Urban Development and Ecology and the governments of such oil-rich states as Chiapas, Tamulipas, Tabasco, and Veracruz. As additional evidence of improvements achieved since 1983, he cited a reduction by two-thirds in the tetraethyl lead content of gasoline and a halving of the sulfur component of diesel fuel sold in the capital. "Furthermore," he added, "we also clear away oil accidentally spilled on land or in water, we clean up land that has been the victim of pollution, restore its natural environmental conditions, we install specially designed equipment in our plants to prevent external spills of harmful wastes, and we instill in every member of the Pemex personnel . . . a clear ecological awareness."[24]

As reassuring as this language seemed, the report remained silent on a November 1984 explosion in a Pemex gas storage and distribution center located in the Mexico City suburb of San Juan Ixhuatepec. Thousands of people sustained injuries and more than 452 died in a tragedy that, according to local residents, the monopoly could have prevented by acting promptly when warned of gas leaks in an area afflicted by previous accidents.[25] Soon after the first of at least five blasts were heard, Beteta completely exculpated the monopoly—a statement greeted by anger, suspicion, and such spray-painted epithets on nearby walls as: "November 19—day of blood," "Pemex guilty—assassins," and "Beteta—Don't look for the guilty—Pemex exists and is close at hand." As criticism mounted, Beteta pledged that Pemex would face up to any responsibility established by "competent authorities."[26]

Moral Renovation

At first blush, it appeared that moral renovation might seriously challenge corruption in the oil sector. A number of former Pemex officials were charged with crimes, and in mid-1983 Mexico's attorney general accused Díaz Serrano of embezzling $34 million in connection with the purchase of two Belgian-built natural gas tankers. Following a stint as ambassador to the Soviet Union, the former director-general had become a senator, and his congressional colleagues voted unanimously to strip him of his legislative immunity. This act foreshadowed a prolonged incarceration in Mexico City's Reclusorio Sur prison before, in May 1987, Díaz Serrano was sentenced to ten years in prison and fined $54 million. Upon being sentenced, he claimed that he was a victim of the "hypocritical moral renovation." "I had the audacity to aspire to the Presidency of the Republic and now I am suffering the consequences of the illogical Mexican political process," he added.[27] Soon after his appointment, Beteta boldly confronted the venal oil workers' union, vowing that "no immoral deals will be made in the shadow of the largest national industry" while he was director-general. He also stressed that the "industry belongs, not to the petroleum workers alone, but to the nation as a whole. Intelligent management and proper use of our resources must favor Mexico, and our basic objective is to serve the community and not to benefit the individual."[28] Such language drew recriminatory snarls from Joaquín Hernández Galicia, the union's de facto chieftain known as La Quina, and Senator Salvador Barragán Camacho, his chief lieutenant and the SRTPRM secretary-general between 1980 and 1984. Increasingly, Beteta and other de la Madrid stalwarts opted for isolated skirmishes instead of unconditional warfare against the powerful union—a strategy doomed to failure because it infuriated SRTPRM bosses without delivering the coup de grâce.

For example, in the spring of 1983 Pemex officials helped remove *quinista* officers accused of misconduct from the leadership of Local 43, a small,

750-employee maintenance unit located in Mexico City, even though Beteta later proclaimed with a straight face: "Union problems should be resolved by their own militants."[29] Syrupy assurances notwithstanding, a more threatening blow to the bulging exchequer of a union that takes in an estimated 2 billion pesos annually[30] came in early 1984. At that time, the Ministry of Planning and Budget promulgated regulations to terminate the subcontracting or selling of public contracts to third parties,[31] a longstanding practice that the SRTPRM and phony "letterhead companies" set up by its leaders had used to earn tens of millions of dollars.[32] Henceforth, the government would award contracts through competitive bidding to entities capable of completing the stipulated work.

Even though the new rules, designed to save the government $1 billion in 1984, would cost the oil workers' union some $165 million, Barragán Camacho complemented private fury over the change with public acceptance: "The secretariat [SPP] once again has adopted a plan that benefits the higher interests of the country," the secretary-general told reporters.[33] What explains this docility? Why didn't union leaders retaliate with strike threats or actual work stoppages? First, they were caught off guard, having just six months before had their right to control onshore drilling expanded from 40 percent to 50 percent—a contract provision later annulled by executive fiat. Second, rather than single out the *petroleros*, the new guidelines applied to all unions as well as businesses. "They can't say anything,"one official told the *New York Times*. "Their backs are against the wall. We said we wanted to clean up the union gradually, without draconian measures."[34] Third, provoking a confrontation with the government over $165 million—especially if the dispute escalated to deploying Mexico's army against the oil workers—would place in jeopardy the billions of dollars of wealth enjoyed by the SRTPRM, not to mention the legal and illegal perquisites savored by its leaders. Legally, the union collected 2.5 percent of its members' wages in dues, a 2 percent commission for "social works" from all contracts handled by outside firms, and the income from businesses that range from fish farms to construction companies.[35] It also enjoyed a number of contracts for maintenance and other functions with the state monopoly. For instance, in 1986 Pemex gave the union approximately $30 per month for each full-time worker. This money was to be used to redeem the food coupon or *vale* that Pemex issues to its employees as a fringe benefit. These coupons could be redeemed only in union stores. Fourth, the union's bucking de la Madrid and the PRI could have diminished the considerable political influence that it wields in cities like Poza Rica, Cuidad Madero, Tampico, and Salamanca, and states like Tamaulipas and Veracruz which are SRTPRM strongholds. Moreover, in 1986 the union boasted one federal senator and five deputies in Congress. Fifth, overt threats to reduce oil output would have triggered public outrage against union leaders, fanned by a government-manipulated press, and could even have driven a wedge between the SRTPRM and the Confederation of Mexican Workers whose leadership cooperated in efforts to promote economic recovery during de la Madrid's administration. Finally, it is possible that the union acquiesced in this reform as a quid pro quo for the government's turning a blind eye to crimes allegedly committed by SRTPRM leaders.[36]

While spurning defiance that would thrust it into open opposition to the chief executive's popular anticorruption initiative, the union filled newspaper columns and the airwaves with self-righteous statements about putting its own house in order. "As anywhere else, we have thieves here, incrusted like a tumor that is difficult to remove," stated union leaders. "But we fight against them and we shall continue fighting with all possible means."[37] Typical of this propaganda barrage was an August 1983 memorandum to Beteta in which Hernández Galicia and Barragán Camacho extended to Petróleos Mexicanos an invitation to inspect the syndicate's accounts and facilities. They concluded their missive as follows: "We are aware that one of the goals of the new administration is to make a new Pemex. For a long time, we have been struggling for the same thing and also for a union more revolutionary, more efficient, and more nationalistic—because we have already demonstrated our patriotism innumerable times."[38]

Meanwhile, two books written in luminous ink circulated about the diminutive, soft-spoken La Quina, depicting his childhood poverty, his personal struggles, and his wrenching sacrifices in behalf of the *petroleros* in particular and the Mexican people in general.[39] As damning as these books were laudatory was Héctor Aguilar Camín's novel, *Morir en el golfo,* which attributed to a Quinaesque character, Lázaro Pizarro, the treachery, cruelty, and greed that characterizes the SRTPRM and its chief.[40] Similarly, two recordings were released featuring romantic songs inspired by the union leader's life, revealing his sensitivity and simple and popular spirit.[41] Also emphasizing his virtues were magazine articles such as "La Quina visto por La Quina," published in the weekly *Siempre!* Written by Luis Suárez, known for his interviews of political personages, this five-page photograph-filled encomium portrayed the ruthless strongman as a smiling, loving, playful, grandfather, who was proud of such virtues as loyalty, work, and honor.[42]

This promotional ballyhoo coincided with accusations by Héctor García Hernández, better known as El Trampas ("the trickster"), that La Quina and Barragán headed a "criminal band" that diverted millions of dollars in union funds to their personal use. El Trampas rose from chauffeuring La Quina to becoming an immensely wealthy entrepreneur and newspaper editor in Coatzacoalcos, where García Hernández directed union operations while serving as the organization's national secretary of education and social security. El Trampas fired his salvo in the form of a letter to de la Madrid, dispatched from McAllen, Texas, where he owned a luxurious condominium, one of his many residences. Soon after the letter became public, three armed union toughs showed up on García Hernández's doorstep and whisked him across the border to La Quina's home in Ciudad Madero where a semiofficial tribunal was held before authorities imprisoned El Trampas in Mexico City on charges of defrauding the SRTPRM of $6 million.[43]

The wily union leaders also attempted to deflect attention from their own Augean stable by condemning practices in Pemex's executive suite. In particular, Hernández Galicia and Barragán Camacho excoriated the creation of several hundred new administrative entities and the hiring of some 3,000 *empleados de confiaza* by Beteta, who was scorned as a petroleum "dilettante." Even though the number of subdirectorates was reduced from nine to seven, there were 321 new coordinating offices (*coordinaciones*), departments (*gerencias*), divisions (*subgerencias*), superintendencies or areas, (*superintendencias*), bureaus (*unidades*), and departments (*departamentos*) distributed as follows throughout the sprawling corporation, as depicted in table 7.[44]

A striking example of bureaucratic expansion took place in the foreign sales

TABLE 7
Changes in Number of Empleados de Confianza *and Coordinating Offices, 1982–1984*

Subdirectorate	Nov. 30, 1982	May 31, 1994	Increase	New Vacancies	Total New Positions	% Increase	No. of New Coordinating Offices
Planning and coordination	187	608	421	248	669	358	17
Primary production	672	674	2	107	109	16	12
Industrial transformation	457	548	91	259	350	76	0
Commerce	799	1,506	707	435	1,142	243	208
Finance	936	1,104	168	242	410	44	40
Technical administration	1,587	1,759	172	109	281	18	31
Projects and construction	539	629	90	28	118	22	10
Administration (dirección general)	51	55	4	12	16	31	3
Total	5,228	6,883	1,655	1,440	3,095	60	321

Source: Material from Local 34 of the SRTPRM, August 16, 1984, published in Angelina Alonso and Robert López, *El sindicato de trabajadores petróleros y sus relaciones con pemex y el estado, 1970–1985* (Mexico City: El Colegio de México, 1986).

branch of the commercial sector. Formerly, there was a single Foreign Commerce Department under the Commercial Subdirectorate. Beteta's reorganization transformed this agency into a Coordinating Office for International Commerce, under which emerged five departments with responsibility for administration, petroleum products marketing, petrochemical products marketing, crude exports, and the analysis and evaluation of international markets. The Subdirectorate for Planning and Coordination underwent an expansion that was second only to the Commercial Subdirectorate.

Pemex insiders have derided the 3,000 new employees as *pitufos* or "smurfs" because of their large numbers and the mystery clouding their origins. Veteran union and nonunion employees condemned the newcomers for their lack of background in the industry, the bureaucratic confusion surrounding their arrival, and the sprouting of the ever denser bureaucratic thicket which they inhabit. Contributing to their unpopularity was the air of superiority often exhibited by the *pitufos*, who were placed in high-level posts to which career professionals had aspired for ten to fifteen years. Moreover, detractors have deprecated unusual expenditures by Beteta's team that prided itself publicly on both "republican austerity" and the prudent use of resources. Among these outlays were the continuation of so-called confidential lists or *nóminas confidenciales* containing names of upper-level executives who received additional monthly stipends of 23,500 to 118,500 pesos. The regular salary and fringe benefits of a manager totaled 208,000 pesos per month; inclusion on the *nómina* raised his overall compensation to 326,500.[45]

Other questionable practices for a firm intent on belt tightening included (1) "special gratifications" or bonuses ranging from 75,000 to 250,000 pesos paid to four engineers who helped extinguish a fire in Nicaragua's petroleum facilities,[46] (2) payment of a manager's wife's 198,006-peso hospital bill,[47] (3) the extensive personal use by company executives of Pemex-rented automobiles,[48] (4) detailing full-time Pemex employees to the PRI and to the friendly opposition Popular Socialist party,[49] (5) renting office space in the Polanco section of Mexico City despite the recent opening of the fifty-two-floor Torre Pemex,[50] (6) authorizing excessive payments for the lavish remodeling of offices, including the director-general's,[51] (7) awarding large contracts on a noncompetitive basis,[52] and (8) forgiving substantial debts owed to the state corporation by the Oil Workers' Union, to which gifts and loans were regularly made in addition to payments stipulated by the collective labor contract.[53]

Corrupt practices also continued within the SRTPRM, though at a somewhat reduced level because of the depression, the worst since the 1930s, that afflicted Mexico in 1982. In that year, the economy stagnated, and in 1983 the GDP shrank 5.3 percent. Possibly these critical conditions, combined with mounting union-related violence, prompted the government to extend an olive branch to leaders of the labor colossus lest the zealous pursuit of moral renovation spark strikes and sabotage in the economy's most important sector. Mexican policymakers recalled the pivotal role of striking Iranian oil workers in toppling the shah from his Peacock Throne. "The only way to cow the union is to have a coup d'état. You actually have to use the army or don't try at all," said Rogelio Ramírez de la O, the director-general of Ecanal, an economic consulting firm. "Maintaining the social peace is a creed in Pemex," according to José Luis García-Luna Hernández, director of the Mexican Petroleum Institute. "You can't push the union to the point of having an explosion."[54] Such an explosion seemed possible because, according to union dissidents, La Quina's forces had trained teams of saboteurs already in place at key facilities.[55]

Beteta initiated the rapprochement with the SRTPRM on May 27, 1984, when he addressed a special assembly of Local 1, in Ciudad Madero, La Quina's hometown. In a speech that began, "Friend Joaquín, friend Chava" (Barragán Camacho's nickname), he launched what critics called the "official sanctification" of Hernández Galicia. The director-general, whom the SRTPRM chieftains treated with crude disdain, praised the "pristine patriotism" evinced by union leaders who "have understood the necessity to support without reservation . . . the policy of Miguel de la Madrid." Furthermore, he said, "I am certain that this is not only the beginning, but the ratification of a commitment, of a relation increasingly better, increasingly solid, increasingly sincere and increasingly full of true fondness between the syndicate and the administration of Petróleos Mexicanos."[56]

Within weeks of Beteta's comments, de la Madrid added his own hyperbole. In a speech delivered at still another special assembly of Local 1, held on July 4, 1984, the chief executive stated: "I wish to recognize the petroleum union for its tenacity, its effort, its sense of loyalty and its commitment to make Petróleos Mexicanos, each day more, a model of Mexican enterprise; a model of honorability, a model of honesty . . . under the direction of its union."[57]

Following the president's Munich-like appeasement of the SRTPRM, its leaders lauded "our good friend" de la Madrid even as they escalated their verbal assaults on Beteta. The most vitriolic attack occurred on June 1, 1986, when La Quina ridiculed Pemex's "arrogant" director-general as a "fourth-rate banker," who was "squandering" Mexico's hydrocarbon wealth by selling oil for $9 to $12 per barrel to Japan and whose disregard for maintenance imperiled both the lives of hardworking men and women and a continued supply of gasoline and other products.[58] Earlier, the union alleged that 40 percent of Mexico's refineries might explode due to Pemex's neglect and deterioration. Indeed, Pemex had been hit by a series of accidents and explosions in recent months, and the possibility of sabotage led to the deployment of army units around important installations usually protected by elements of Pemex's own 3,000-man security force.[59]

Why did the union lash out so violently against Beteta? No doubt there was concern about safety, because maintenance and spare parts had been casualties of the retrenchment imposed on Pemex by a government anxious to reduce a growing budget deficit. Further, the ethos of many Pemex decision makers recruited during the Díaz Serrano period was to produce oil at any cost. And, since 1975, union membership has been required of almost all professionals, some of whom obtained their positions through payoffs to union leaders, instead of presenting compelling credentials. Even more important, the downturn in energy earnings, coupled with reforms backed by Beteta, hurt the union financially. A case in point was Beteta's insistence on shipping ever higher volumes of crude oil and products in Pemex tankers rather than those owned by either Petroflota, a union company, or Sergio Bolaños, the SRTPRM's immensely weathy financial adviser. A further setback would result if the director-general followed through on a pledge to shrink Pemex's bloated labor force. He knew that Venezuela with 44,000 oil workers produced almost as much oil as Mexico, which boasted more than four times as many *petroleros;* indeed, under Beteta the Pemex labor force had swollen to 187,649 men and women, 176,734 of whom belonged to the union.[60] Rumors abounded in the Torre Pemex of the director-general's plan to break up the monopoly, along functional lines, into four or five corporations—a move that would have severely weakened the union.

La Quina's worst headache was that Beteta might become Mexico's next chief executive and deepen the reforms begun during de la Madrid's *sexenio.* In fact, the fusillade directed at Beteta shattered his prospects for becoming president—a fact that became evident in April 1986 when he failed to be named energy secretary upon Labastida's nomination for the governorship of Sinaloa state. Ten months later, Beteta left Pemex to become the PRI's gubernatorial candidate in Mexico state, the country's most populous and industrialized. Since chief executives traditionally come from the cabinet, the new post did not improve Beteta's chances for the presidency; rather, it marked a prestigious reward for de la Madrid's close and loyal friend, who had been publicly humiliated by the union leaders. Nonetheless, the reforms championed by Pemex's former director-general set the stage for additional curbs on the SRTPRM's still bountiful opportunities for enrichment.

And Hernández Galicia made no secret of his contempt for Salinas as the PRI's nominee. After all the former SPP secretary, a Beteta ally, had pushed through the 1984 bidding requirement for government contracts that the Oil Workers' Union found so odious. Even as he begrudgingly pledged his support to Salinas, La Quina (1) assured the technically oriented PRI nominee that he was not the *petroleros'* first choice, (2) publicly chided him for fashioning an austerity program deemed harmful to workers, (3) flamboyantly made financial gifts to the CTM and other recipients to demonstrate the union's wealth and power, (4) extended the SRTPRM's secretary-general's term from three to six years to coincide with the *sexenio*, (5) placed loyalist Barragán Camacho in the number one union post, and (6) had himself elected alternate secretary-general. These moves, which came in early November 1987, revealed Hernández Galicia's determination to fight any Salinas-inspired modernization of the petroleum sector that might imperil the union's dominance. Salinas's administration could witness a showdown between the president and La Quina, men who represent diametrically different philosophies and styles in Mexico's complex political system.

Meanwhile, Francisco J. Rojas, who held the position of comptroller-general before succeeding Beteta, seemed more interested in paring the number of *pitufos* than in confronting La Quina. As of mid-1987, he had eliminated approximately 1,200 positions, while "compacting" the size of the Commercial, Financial, and Planning Subdirectorates. Rojas also stressed his commitment to safeguarding the environment, improving maintenance, and reducing the accident rate. Both the monopoly and the SRTPRM contributed to a 15 billion peso fund created to address emergency maintenance needs identified by a tripartite Pemex-union-government commission.[61]

The Diplomacy of the New Pemex

The new Pemex adhered to many of its predecessor's nationalistic totems and taboos with respect to selling oil abroad. It continued the policy, begun under López Portillo, of limiting Pemex's sales to any one country to 50 percent of total exports. While no nation was mentioned, this guideline clearly targeted the United States, by far the largest oil customer of Mexico, which—in turn—surpassed Saudi Arabia between 1983 and 1985 as the number one external supplier of the U.S. market. On the one hand, exceeding the 50 percent threshold amid a worldwide shortage might excite foreign designs on Mexico's oil industry, which had been controlled by U.S. and European firms until the celebrated 1938 expropriation. On the other hand, excessive reliance on a single market could increase Mexico's vulnerability to the imposition of special taxes or import restrictions imposed during periods of surplus. Concerns rooted in nationalism and security also underlay continued insistence that no country should depend on Mexico for more than one-fifth of its crude imports—a maximum from which Israel and San José Accord beneficiaries were exempted.

Pursuant to past conventions, the government continued to set export prices instead of adopting market charges for the 1.5 to 1.7 million bpd shipped abroad under the five-year plan. Nationalism suffused the enduring policy of selling only to final consumers/refiners to prevent retrading and, perhaps, profiteering at Mexico's expense. After all, the Spanish despoiled Mexico of its gold and silver in the sixteenth, seventeenth, and eighteenth centuries, while North American and British entrepreneurs made handsome profits from silver, oil, and other minerals in the nineteenth and early twentieth centuries. Hence, contemporary Mexican leaders were determined to prevent foreign exploitation of their country's rich patrimony.

Persisting unwillingness to sell on the noncontract or spot market derived from the difficulty that fluctuating prices posed for long-range economic planning keyed to oil-generated revenues. Inadequate storage capacity was also a consideration. In addition, it was doubtful that Pemex boasted personnel sufficiently experienced to wheel and deal as effectively as required of successful spot market traders. Of course, the light beamed on corruption in recent years might have made Pemex commercial specialists resist any spot selling because failure to attain the maximum return on even a single transaction could have unleashed accusations of malfeasance.

Beteta's most publicized institutional innovation in foreign matters was the formation of the Foreign Oil Trade Committee (COCEP) in December 1982 to review the international petroleum market and advise Petróleos Mexicanos on pricing Mexican crude. Chaired by the director-general himself, COCEP was also composed of undersecretaries from five ministries (Energy, Mines and Decentralized Industry; Finance; Commerce and Industrial Development; Foreign Relations; and Planning and Budget as well as representatives of Pemex and the Central Bank of Mexico). Such broad-based membership favored the consideration of a variety of viewpoints and interests before price changes were announced. Thus the presence of COCEP encouraged integrated policymaking in contradistinction to the freelance approach characteristic of Díaz Serrano's highly personalized reign. Collective advice spread responsibility for pricing moves among seven government entities.

At first COCEP's monthly meetings provided a forum for ratifying policy outcomes endorsed by Pemex. Beteta chaired the sessions, and Pemex's representative had more information on sales and market conditions than other participants. After 1983 the deliberations of COCEP changed. As the members learned more about the domestic and international energy picture, representatives of the ministries and the Bank of Mexico became less deferential toward the state monopoly. Especially diligent in gathering information were the Ministries of Commerce and Energy. Their officials sought and received data from Pemex's New York office, which carefully monitors U.S. and worldwide energy matters. That the Energy Ministry viewed energy developments from a broader perspective than Pemex, which historically has concentrated on production, refining, and export objectives, sometimes generated friction between the two agencies. Increased knowledge by COCEP members expanded the number of questions raised and precipitated more intense discussions at its meetings. Pemex—in part because of Beteta's close and longstanding ties with de la Madrid—still prevailed on most important issues.

COCEP-debated issues such as price changes could be elevated to the "economic cabinet," a specially created informal grouping that—under de la Madrid—embraced the heads of the government bodies composing COCEP plus the comptroller-general and the secretaries of labor and interior. The president chaired the intra-agency body whose counsel he would more likely follow in case of a conflict with COCEP. Nevertheless, relations between the "economic cabinet" and COCEP traveled on a two-way street inasmuch as recommendations of the former might be referred to the latter and vice versa.

Pemex, the Central Bank of Mexico, and the Finance and Commerce Ministries abounded with liberal-rationalists who were attuned to market forces—a sensitivity increasingly evident in COCEP, owing both to the advocacy by the four agencies and to competitive forces accentuated by the oil glut. Therefore, they urged flexibility in pricing to protect Mexico from suffering a severe loss in earnings like the one that occurred in mid-1981 when López Portillo, to prevent undercutting OPEC, failed to act astutely when a seller's market shifted in favor of buyers. As will be discussed in chapter 6, nationalist-populists in the Ministry of Foreign Relations and, to a slightly lesser degree, in the Energy Ministry, championed collaboration with the cartel in lieu of prompt responses to supply and demand factors. Those advocates of market forces underscored the need to husband nonrenewable resources that could be exhausted in a generation, with Mexico left holding a bag full of obligations to international banks. As Jorge Eduardo Navarrete, the undersecretary for economic affairs within SRE, expressed it: "You have some people saying we should exploit the reserves immediately and quickly, while others say we should go easy. I, for one, believe we are in the oil business from here to the next century, not just to take advantage of short-term windows of opportunity."[62]

Developments between fall 1984 and spring 1985 revealed the mounting influence of the market-oriented pragmatists over advocates of OPEC solidarity. Effective November 1, 1984, Mexico cut exports by 100,000 bpd to assist the cartel that, although still plagued by cheating, had just slashed output quotas from 17.5 million to 16 million bpd to dissipate the worldwide surfeit of crude.

Three months later, persisting market weakness prompted nine OPEC members to agree to what was only the second price cut in the organization's twenty-five-year history. Among the reductions decided upon at the January 1985 Geneva session was lowering the price of Arabian Light, the industry's benchmark crude, from $29 to $28 per barrel. On February 4 Mexico—furious over the readiness of unscrupulous exporters to exceed their quotas by as much as 50 percent—answered the OPEC move by restoring its export level to 1.5 million bpd and trimming its charge for the Isthmus variety to $27.75, 25 cents below that of the comparable Arabian Light. The price adjustment decreased export earnings by more than $300 million and required de la Madrid to implement new austerity measures.

Neither OPEC nor non-OPEC price cuts soaked up the surplus, and cheating by sellers continued—with an estimated 80 percent of world oil transactions in May and June taking place at prices "substantially below" official OPEC rates.[63] Nevertheless, Sepúlveda and Labastida, secretaries of foreign relations and energy, respectively, urged the maintenance of Pemex prices to buttress OPEC's effort, undermined by individual members, to stabilize a sagging market. Apparently de la Madrid, preoccupied during much of May with an impending visit to Europe, accepted what turned out to be extremely expensive advice. Retention of the February prices found Mexican exports plummeting to 683,000 bpd in June 1985, one-half of the target, as clients patronized either the spot market or exporters offering discounts and other inducements.

Eventually, Mexico acted. On June 17, 1985, it reduced the price of the heavier Maya crude from $25.50 to $24, retroactive to the first of the month. But this was a classic case of doing too little, too late. To win back customers, Pemex announced a broader cut on July 10 as it decreased the price of Isthmus by $1.24 to $26.51 and that of Maya by 77 cents to $23.23. A senior OPEC official labeled the move, taken in anticipation of any cartel action, as a "very serious disappointment."[64]

Particularly strong market resistance in Europe and Asia threatened Mexico's geographical diversification policy. To prevent U.S. purchases from soaring above the 50 percent mark, Pemex offered lower prices to its European and Far East customers through a three-tier pricing policy.

The episode left the pragmatists clearly in the ascendancy, their attentiveness to market forces vindicated as July 1985 sales climbed to 1.48 million bpd. For his part, Labastida assumed a notably lower profile until selected for the Sinaloa governorship, while a mid-1985 measure devised to shave the budget deficit eliminated, among other agencies, the Ministry of Foreign Relations' Undersecretariat for Economic Affairs. Navarrete, the head of this entity, had militantly backed a united front with OPEC producers.

Nevertheless, exports declined to 1.1 million bpd in early 1986 in the wake of plummeting world prices, as Mexico struggled to devise a pricing mechanism attractive to customers being wooed by Saudi Arabia and several other producers who discounted their crude, by linking its charge to the value of petroleum products refined from each barrel. Among other considerations, nationalism foreclosed Pemex's adopting this "netback" approach. Turning over pricing matters to foreign refiners was anathema to a government for which the 1938 expropriation of foreign petroleum corporations remained a red-letter event in its revolutionary tradition. COCEP experimented with monthly, twice monthly, and daily pricing before announcing on May 28, 1986, that it had devised a competitive pricing scheme that had boosted exports to almost 1.4 million bpd during the previous six to eight weeks. Under this mechanism Mexico's FOB prices for its two crude oils adjusted daily in relation to the soft market prices for a group similar crudes in two of its three market areas: the Western Hemisphere, and Europe. For example, West Texas Intermediate, West Texas Sour, and Alaska North Slope graced the basket of relevant crudes for the Western Hemisphere where in September 1986, Saudi Arabia supplanted Mexico as the number one supplier to U.S. customers after Venezuela (Venezuela remained the largest supplier during 1987). Also, fuel oil costs were factored into the formula, as was a constant and destination fee for Maya and Isthmus, respectively. Lack of Japanese enthusiasm for the market basket approach led to negotiated retroactive prices in the Far East keyed to charges for Dubai Light crude. Whether Mexico, aided by Pemex's computer models, could keep abreast of rapidly changing conditions remained to be seen. Still, the ever greater influence of supply-and-demand in setting prices was quite impressive as export earnings for the first nine months of 1987 ($6.4 billion) exceeded the 1986 total ($6.1 billion). This influence, dramatic in both domestic and foreign sales of the economy's most vital sector, reinforced the importance of market forces as the country embarked upon a liberalized, export-focused development model. The sheer size of the "new Pemex" meant that its internal and external policies would radiate throughout the economy. In the words of Beteta: "Perhaps no [institution] has as great an importance and capacity to pull the country out of its economic problems as Petróleos Mexicanos."[65]

United States–Mexican Relations

On May 12, 1986, Senator Jesse A. Helms (R-N.C.) gaveled to order the first of a series of hearings on Mexico, conducted by the Subcommittee on Western Hemispheric Affairs of the Senate Foreign Relations Committee. The narcotics issue dominated the two-day session as William von Rabb, commissioner of the U.S. Customs Service, excoriated drug trafficking in Mexico, which— in 1985—reportedly supplied 30 percent of all cocaine entering the United States and surpassed Colombia and Asia to become the largest conduit for marijuana and heroin.[1] Von Rabb told of "engrained" official corruption "up and down the ladder" of Mexico's law enforcement establishment, claimed that the governor of Sonora owned four ranches that produced marijuana and opium poppies, and implied that de la Madrid's relatives were involved in the narcotics business. Similarly, Elliott Abrams, assistant secretary of state for inter-American affairs, claimed that Mexican corruption related to the drug trade was so pervasive that "in a number of years, the traffickers' influence on government may be so strong that it . . . [will be] hard or impossible to root out." Abrams warned that Mexican authorities "have got to get organized to stop this before it gets too late, and it can get too late. . . . We have told the Mexicans in no uncertain terms that we are deeply troubled by widespread drug-related corruption."[2]

Mexicans greeted the Helms hearings as the political version of fingernails clawing a blackboard. For them, the soft-spoken, bespectacled senator from North Carolina seemed to embody the spirits of General Winfield Scott, General John Pershing, Ambassador Henry Lane Wilson, and other North Americans who had meddled egregiously in their nation's affairs. "Let's stick it to the gringos!" was the sanitized translation of the collective response that rippled through universities, union halls, and government agencies.

In an official protest note presented to the State Department, Ambassador Jorge Espinosa de los Reyes decried the intervention in Mexico's internal affairs and strongly rejected "the accusations and calumnies pronounced against Mexico in the hearings"; on May 13 *El Nacional,* the government-run newspaper, called the hearings "intolerable" and said that they "muddy the cordial relations between our countries." The Foreign Ministry referred to "slanderous accusations" made by "minor functionaries and foreign legislators."[3]

In the aftermath of the first subcommittee session, a Molotov cocktail exploded outside the U.S. consulate in Guadalajara; tens of thousands of demonstrators, led by Porfirio Muñoz Ledo, Mexico's former U.N. ambassador, María de Echeverría, wife of the ex-president, and party notables, marched into Mexico City's Zócalo to repudiate U.S. intervention in their affairs; the leader of the Communist-dominated United Socialist party of Mexico (PSUM) called for a moratorium on the external debt; and the president of the Mexican Labor Congress announced that the thirty-three affiliates of his organization would support unconditionally any government decision to stop paying foreign creditors. The apology of Attorney General Edwin Meese III to his Mexican counterpart did not assuage the widespread anger.[4]

In fact, the resentment spread as leading U.S. newspapers ran features on our "neighbor in distress," and a report of the Central Intelligence Agency, according to a reliable leak, estimated that there was a 20 percent chance that the Mexican political system would break down completely within five years.[5] The appearance in the United States of potboilers such as Sol Sander's *Mexico: Chaos on Our Doorstep* exacerbated the indignation.

The shrill outcry against Washington coincided with an even more notable event in the Mexican capital—the month-long World Cup championship. At the May 31, 1986, opening ceremonies, the largely middle-class crowd of 100,000 gave de la Madrid Mexico's sibilant version of the Bronx cheer three times: when he entered the stadium, when an international soccer official praised him, and when he attempted to make brief welcoming remarks.

It seemed possible that the Helms hearings coupled with the stadium rebuke might convert the president's studied pragmatism into demagogic opportunism. After all, as a technocrat who lacked a feel for the popular pulse, he might have concluded that Uncle Sam was out to get him, while pursuing the role of a Responsible Debtor—that is, staying the course with an unevenly applied, but impressive, austerity and structural reform program—would only worsen his standing at home.[6]

The day after the soccer incident, Mexico's offended *jefe máximo* flew to Sonora where he vowed to "defend Mexico's sovereignty and independence" by preserving every comma in his country's highly nationalistic investment statute.[7] This law restricts foreign investors to minority ownership in ventures unless the government grants an exception—as occurred in 1985 when International Business Machines agreed to invest $91 million in microcomputer operations. Revising the so-called Mexicanization requirement would have signaled to the domestic and international business community greater congeniality to private investment in a country where the state generates or controls more than 60 percent of the Gross Domestic Product. In the face of formidable challenges near the end of their mandates, de la Madrid's immediate predecessors embarked upon populist initiatives to ingratiate themselves with the masses. In both cases, their actions—Echeverría's limited land redistribution in 1976 and López Portillo's nationalization of the banking system in 1982— aroused short-term approval at the cost of alienating the private sector, impelling capital flight, and sharpening economic problems.

At first it appeared that Silva Herzog's removal as finance minister in June 1986 might portend an impulsive move such as reneging on foreign debt payments. However, de la Madrid quickly filled the position with Gustavo Petricioli Iturbide, a fifty-seven-year-old economist who belonged to the president's *camarilla* and whose orthodox economic views coincided with Silva Herzog's. Petricioli proved to be a faithful and unquestioning executor of presidential policies, for he lacked his predecessor's magnetic personality, worldwide reputation, and international contacts. Thus, despite the churning on the surface of bilateral relations, deeper currents of pragmatism prevailed —especially with respect to mutual consultations on energy questions and natural gas policy. Nonetheless, despite extraordinary efforts by Pemex, Mexico has yet to demonstrate effectiveness in exerting influence in Washington so important both to preventing Helms-style "Mexico bashing" and creating an environment propitious to advancing the interests of a Responsible Debtor.

Bilateral Energy Consultation Group

The Bilateral Energy Consultation Group (BECG) has fostered a more businesslike atmosphere in U.S.-Mexican energy relations, thereby contributing to Mexico's image as a Responsible Debtor. Modeled after a highly successful joint U.S.-Canadian commission, the BECG was created in 1982 as a forum for discussions between U.S. and Mexican specialists. Generally speaking, such discussions have taken place twice annually since their inception, with meetings held alternately in Mexico City and Washington. The United States government views these informal consultations in a bilateral context. Yet, from the Mexican perspective, the talks fall within the framework of de la Madrid's promotion of a dialogue between producing and consuming countries, conceived to contribute to market stability without jeopardizing Mexico's independence and sovereignty.[8]

Originally, an undersecretary in the Department of Energy (DOE) chaired the U.S. delegation; nevertheless, within a few months after the BECG's establishment, a State Department official had been named co-chairman, and the two departments handled arrangements for sessions convened in Washington. On the Mexican side, an undersecretary in the Ministry of Energy, Mines, and Decentralized Industries (SEMIP) chaired some sessions alone, others jointly with a Pemex representative. However, by 1986 the Ministry of Energy appeared to be guarding jealously its position of leadership of the Mexican delegation.

Budget considerations mean that the U.S. contingent is small when the BECG meets in the Mexican capital. When the group convened in Hacienda Cocoyoc, south of Mexico City, on April 23–24, 1987, the U.S. delegation embraced four DOE representatives and two from the State Department (including the U.S. embassy's petroleum officer). The U.S. delegation is larger when sessions occur in Washington. For instance, an official from the Office of the U.S. Trade Representative may speak on legislative developments on the energy front, while DOE experts address subjects as varied as conservation and nuclear energy. Typically, these specialists present a briefing, answer questions, and leave. Ten or more officials, roughly divided between SEMIP and Pemex, are the core of the Mexican delegation for meetings held at home.

Only a small group of Mexicans, composed of men and women from SEMIP and Pemex, attend Washington sessions. Also participating in the April 1986 meeting was José Luis Bernal, first secretary for economic affairs in the Mexican embassy. Usually, Alfredo Gutiérrez Kirchner, Pemex's New York and Washington general representative who also serves as minister-counselor for energy in Mexico's embassy, takes part in sessions held in both countries.

The sessions generally begin with an analysis, often provided by the Department of Energy, of the world energy picture, stressing supply, consumption, inventories, and prices. Next the participants take turns describing energy developments within their respective nations—with Mexico particularly interested in the analysis of energy markets by DOE's technical experts and legislative initiatives such as the superfund tax levied on all oil imports and a proposed oil import fee. The United States has shown greatest concern about Mexico's relations with OPEC, its export ceiling, its exploration and development activities in the face of severe budgetary constraints, and its provision of crude oil to the U.S. Strategic Petroleum Reserve (SPR). Mexico's delegation took advantage of the November 19, 1984, bilateral conference to outline its country's recently announced National Energy Program. Later, Mexico's formula for setting export prices became the center of attention. Finally, the participants have concentrated on such matters of topical interest as U.S. administrative actions affecting gasoline and other light product imports, and the friendly and forced mergers involving U.S. oil companies.[9]

BECG sessions are conducive to a frank exchange of ideas and expressions of concern about outstanding bilateral problems. Agendas are flexible; the preparation of position papers is discouraged; media attention is minimal; the participants issue no communiqué upon adjournment; and the goal of the specialists is not to reach decisions but to inform each other—a practice that discourages inflammatory speeches and political posturing. Although Beteta and Ambassador John Gavin took part in a "working luncheon" at the June 1985 meeting, the two governments have strongly deemphasized protocol in favor of substance with discussions carried on by officials below the deputy assistant secretary level.

Above all, the group has provided Mexico with a more realistic picture of U.S. and world energy conditions. The DOE's supply of technical information is important in this regard, particularly to SEMIP which, in terms of data, is at a conspicuous disadvantage vis-à-vis Pemex, for the latter boasts a sophisticated computer capability, a New York office that collects data on the international oil market, and an impressive staff of analysts. Mexico's access to expertise through the BECG has proven especially valuable as the prices obtained for Pemex crudes halved during the first quarter of 1986 only to recover later in the year.[10]

In addition, U.S. officials have sought to allay misgivings expressed by some of their Mexican counterparts, who—for instance—feared that increased mergers in the U.S. oil sector might foreshadow the reemergence of the "seven sisters"; namely, the domination of the oil industry by a small collection of transnational behemoths. Discussions about pending legislation have helped illuminate the complex, decentralized character of the U.S. political system, thereby challenging a widely held impression among Mexicans that it was, like their own government, markedly centralized and hierarchical. The ever more informal meetings have enabled the U.S. and Mexican officials to become acquainted with each other as individuals, not simply as agents of another country. One participant, who asked to remain anonymous, expressed satisfaction over the increasing signs of camaraderie evident at the meetings and social events surrounding them. Holding the early 1987 session in a resort near Cuernavaca contributed to the relaxed atmosphere. He, like others, praised the consultative mechanism for building confidence between the two countries on energy questions. Clearly, the Mexicans have perceived an enhanced status associated with meeting on equal terms with the United States whose unparalleled information on energy is invaluable to SEMIP and Pemex. The BECG's success was reflected in the fact that the June 1985, April 1986, and April 1987 meetings lasted two days instead of one. Still, the tendency has been to hold fewer meetings, possibly because other assignments have preoccupied the relatively small number of Mexican energy experts. Meanwhile, there appeared to be less pressure on the two governments to meet during a period of lower prices.

Natural Gas Policy

The pragmatism evident in the Bilateral Energy Consultative Group has also characterized Mexico's natural gas export policy during de la Madrid's administration. The gas policy of a Responsible Debtor offered a sharp contrast to Mexico's approach to bilateral gas sales in the 1970s when López Portillo was seeking regional leadership. Following protracted, sometimes acrimonious, negotiations with the United States in 1979, Mexico finally agreed to sell 300 million cubic feet per day (cfd) of gas to its northern neighbor.[11] On January 15, 1980, an engineer turned the valve on a pipeline in northern Mexico to commence the flow to Border Gas, Inc., a Houston-based consortium of six U.S. pipeline companies headed by the Tennessee Gas Transmission Company, a Tenneco subsidiary.[12] Under the contract between Pemex and Border Gas, the price was $3.625 per thousand cubic feet (Mcf)—with Mexico having the right to adjust future prices in accord with either those charged by Canadian exporters or a market basket of five crude oil prices (Venezuelan Tía Juana Medium, Isthmus, Saharan Blend, Arabian Light, and UK Forties).[13]

Within ten weeks after deliveries began, Mexico raised its price to $4.47 per Mcf—an amount equal to that which U.S. firms paid Canadian suppliers. Two increases, matched by Canada, brought the Mexican price to $4.94 per Mcf in 1981.

Developments in the United States directly affected imports. The Natural Gas Policy Act of 1978 called for the phased-in deregulation of ''new gas'' until all price controls would be lifted on January 1, 1985.[14] This deregulation, which began in January 1980, sparked increased prices which, in turn, stimulated domestic exploration and development. For example, the 1980–1981 period witnessed the first increase in the number of gas wells drilled in the lower forty-eight U.S. states since 1967. In addition to raising reserves to 201.73 trillion cubic feet by the end of 1981, the higher prices contributed to declining demand—in part because recession gripped the U.S. economy; in part because some industrial users whose facilities were equipped with dual-fired boilers shifted to relatively cheaper residual fuel oil. Approximately one-quarter of the gas consumed in the United States was burned as a boiler fuel. The development of their own new reserves whetted Canada's interest in preserving its share of the U.S. market. Although its population is just one-third that of Mexico, Canada boasts higher reserves of natural gas: 98 trillion cubic feet compared to Mexico's 76.5 trillion cubic feet as of December 31, 1987.

Mexico and Canada provided less than 5 percent of U.S. natural gas needs;

nevertheless, their prices were well above those commanded by North American companies. Consequently, these domestic producers, unable to sell their entire output, criticized Border Gas for purchasing high-priced imports. Consumer groups also voiced their opposition to this practice.[15]

In March 1983 Border Gas informed Pemex that, effective April 1, it would exercise an option to decrease purchases by 40 percent—to 180 million cfd, the minimum volume allowed under the contract. While not discussing price explicitly before taking this action, Border Gas was clearly signaling its unwillingness to continue paying $4.94 per Mcf. Thus, the presence of a gas bubble and a cut in Canadian prices the month before persuaded Pemex to trim its price to $4.40 per Mcf on May 1. These reductions in quantity and price cost Pemex an estimated $169.1 million during the year, as its natural gas revenues fell from $526.2 million in 1981 to $306 million in 1983.

Even at $4.40 per Mcf, Mexico's price was "still substantially above our weighted average price of gas," said Franklin H. Reed, a Tenneco spokesman. "There's a glut of natural gas and we're taking what we can. The lower price isn't a reason to go out and buy more of that [Mexican] gas," he added.[16] Reed said that Tenneco's average buying price was $3.25 per Mcf before the firm instituted an emergency purchase plan on May 1 designed to lower the average price an additional 30 to 40 cents per Mcf.[17]

Canada responded quickly to the market cues emitted by Tenneco and other transmission companies. In mid-1984 Ottawa announced that, beginning November 1, Canadian companies would be allowed to negotiate natural gas export prices with customers. Moreover, natural gas exporters could offer a discount price of $3.40 per Mcf, or lower, for buyers willing to increase their purchases beyond 50 percent of either their contract quantity or the amount purchased the previous year. According to Energy Minister Jean Chrétien: "We hope to preserve our share of the United States natural gas market, and possibly to increase sales in certain regions."[18] Still, under the new policy, the price of gas sold pursuant to long-term contracts could not fall below the wholesale price for gas marketed under similar provisions in Toronto—$3.10 per Mcf at the time the change was effected. All contracts would be subject to review by the National Energy Board, which would be responsible for monitoring contracts and submitting annual reports to the Department of Energy, Mines, and Resources.

The persisting gas surplus precipitated discussions between Pemex and Border Gas during the summer and fall of 1984. Even though Adrian Lajous Vargas, coordinator for international trade, usually represented Pemex at these meetings, Beteta attended one session. In view of Canadian moves and the January 1985 date for the complete deregulation of "new gas," Pemex realized that it could not adhere to the $4.40 per Mcf figure and hope to continue selling in the glutted North American market. To remain competitive, it was clear that Mexico had to reduce its price to approximately $2.75 per Mcf. Possible "political costs" of such a move compounded the economic unattractiveness of the reduction.[19] Thus, Pemex and Border Gas, which had advised the departments of State and Energy of the probable action, issued a joint press communiqué on October 24, 1984, to announce that "the temporary suspension of sales and purchases is in the best interests of both parties." Border Gas President Robert C. Thomas indicated that the "parties have agreed to meet periodically after November 1 to review market conditions relative to a resumption of gas purchases under the terms of the contract."[20]

Pemex, whose gas exports constituted only a fraction of Canada's, decided that it would be more cost-effective to divert the 180 million cfd of gas to domestic consumption, thereby freeing up some 30,000 bpd of fuel oil for sale abroad. Flexibility was also a consideration. The *gasoducto* meant that all gas exports reached the U.S. market; meanwhile, fuel oil could be sold to any of Pemex's nearly two dozen customers. The earnings from these exports would compensate for the annual loss of several hundred thousand dollars in foreign exchange resulting from the suspension of deliveries to Border Gas.

The 1984 natural gas decision was quite different from actions taken in 1977 when Mexico allowed a memorandum of intent with pipeline companies to expire after the U.S. government failed to approve a purchase price pegged to the BTU equivalent of fuel oil delivered in New York harbor. This expiration ushered in eighteen months of invective between the two countries, highlighted by a stinging toast, quoted earlier, that López Portillo inflicted on Carter in February 1979.

What accounted for the difference? To begin with, Mexico's natural gas reserves had not increased as expected—in large measure because deposits in the northern fields near Monclova proved much smaller than anticipated during the oil boom of the 1970s. This discouraged gas exports as domestic users consumed 95.5 percent of the nation's annual production even before the suspension was announced. Further, although Beteta participated in one negotiating session, middle-level energy specialists guided the discussions that preceded the halt of deliveries to Border Gas. The Mexican officials active in natural gas matters in the 1980s were much better informed about the U.S. market than were their predecessors seven years before. They understood the complexities of an essentially economic problem for which a technical rather than a political solution was indicated. In addition, the conversations between Pemex and Border Gas took place outside the limelight, minimizing the temptation to indulge in political posturing or demagogic statements. Finally, even though Border Gas kept Washington informed of developments, the negotiations occurred strictly between the firm and Pemex. Representatives of the U.S. government watched from the sidelines as an action was taken that required no official sanction, thus reducing both the political content of the discussions and the possibility of a confrontation. Needless to say, an abundance of energy available to the U.S. market militated against stridency on the part of Mexico and other suppliers.

Mexican Lobbying in Washington

The turnabout in natural gas conditions reinforced the fact that Mexicans no longer could take for granted a strong demand for their energy exports within the United States. As discussed earlier, losing the energy trump card contributed to a change in Mexico's international role conception. While welcomed by most public and private financial authorities in the United States, Mexico's shift to a Responsible Debtor emboldened some U.S. decision makers to act differently toward a neighboring country which they viewed as more vulnerable. Indeed, members of both the executive and legislative branches who sought Mexico's good will during the energy crisis of the 1970s began, after the appearance of an oil glut, to mutter about drug trafficking, illegal immigration, unfair trade practices, and other nettlesome bilateral matters. In the case of Senator Helms, who treated Mexico with all the delicacy of Cromwell ruling Ireland, the grousing exploded into a philippic against the Mexican political system that was splashed across the front pages of newspapers in both countries. It became increasingly evident that, as the cliché goes, Mexico needed to win friends and influence people, lest the weakening of the oil weapon expose the country, characterized by an ever more moderate foreign policy, to accelerating attacks in the U.S. capital.

Even though several Mexican firms had lobbied in Washington, the Mexican government had failed to do so in a systematic, assertive, and effective manner. López Portillo's first foreign minister, Santiago García Roel, hired the Washington law firm of A. Lee Fentress not only to represent Mexico's interests but also to report on the activities of the country's ambassador, Hugo Margain. Apart from such Machiavellian antics, what explained this reluctance to assume a higher profile in the United States? To begin with, lobbying meant involvement in the affairs of another country and, as such, disturbed proponents of Mexico's cherished, though inconsistently applied, nonintervention principle. A variation on this theme was the antipathy that the Mexican government felt toward the idea of justifying its actions to anyone, especially to its powerful northern neighbor. Even more troubling to some Mexicans was the possibility that attempts to influence Washington decision makers might be seized upon to justify even greater U.S. involvement in its southern neighbor's domestic affairs where John Gavin, U.S. ambassador from 1981 to 1986, had assumed an extremely high profile. Of course, any appearance of pandering to Uncle Sam may be political suicide for those Mexicans involved. One of the many ironies of bilateral relations is that close association with the neighboring country, considered desirable by most U.S. politicians anxious to court Hispanic-American voters, can be the kiss of death to their Mexican counterparts.

Limited resources and cautious ambassadors contributed to Mexico's unobtrusiveness at a time when scores of other countries actively promoted their interests in the United States. Conversations with a large number of staff members on Capitol Hill revealed that Ambassador Espinosa de los Reyes seldom sought meetings with U.S. senators. He complied graciously with requests for interviews, yet did not take the initiative. The ambassador was not a career diplomat, but a banker who once held the post of Pemex's commercial subdirector. However, his inertness set the tone for an embassy in which key foreign service officers increasingly understand the importance of diplomatic activism. Still, Sepúlveda played a highly scripted role in the cabinet as Mexico's designated advocate for Third World solidarity and anti-U.S. nationalism. Just as departments of the U.S. government appeal to specific interest groups—such as Treasury to bankers, Agriculture to farmers, and Commerce to business—the Foreign Relations Ministry has an important constituency that resonates to the Yankee-baiting in which the secretary and his subordinates sometimes engage. Their attacks on imperialism help nurture the idea that Mexico is adhering to its revolutionary tradition, while deflecting attention from severe economic problems. Such policies enjoy support among vocal elements within the PRI, the mass media, academia, the Mexican Congress, and the labor movement. Men and women with this outlook resisted entanglements with the "imperialist" United States, which they regarded as hostile to the interests of Mexico in particular and of developing nations in general. Indicative of the SRE's special mission was Mexico's U.N. record. As seen in table 8, despite greater pragmatism in its overall foreign policy, Mexico increasingly found itself at odds with the United States in the General Assembly—to the point that there was only a 28.7 percent coincidence in votes in 1986 compared to 66.7 percent for Canada. Among the thirty-four nations of the Western Hemisphere, only Cuba, Nicaragua, Guyana, Trinidad and Tobago, and Peru registered lower figures.

Also discouraging Mexican lobbying in Washington was a fatalistic view that virtually nothing could be done about the anti-Mexican sentiment in the United States and, hence, it was futile to cultivate policymakers in the executive and legislative branches. In early 1986, a theory gained currency among Mexican officials that their country was the target of a U.S.-contrived defamation campaign. In March 1986, two days after a cover story on Mexico appeared in *Newsweek* magazine's international edition, Deputy José Angel Pescador claimed to speak for the PRI when he told his legislative colleagues that it was "possible to suppose" that *Newsweek* was simply "following the line assumed by some foreign publications that consistently have tried to distort the image of Mexico abroad and, at times, to degrade the Mexican people." "The interests behind such commentaries can be deciphered," he added, "and it is possible to uncover a strategy to disseminate information aimed at generating a lack of confidence in ourselves, at dividing us and making us confront each other in sterile conversations that lead nowhere."[21] Similar complaints were made when negative media reports on Mexico coincided with de la Madrid's 1984 and 1986 visits to the United States. The press control exerted by their own government led many Mexican politicans to see Uncle Sam's heavy hand in U.S. media reports critical of Mexico. Further impeding their nation's effectiveness in Washington was a persistent tendency of many Mexican leaders to perceive the U.S. political system as the mirror image of their own authoritarian, presidentially dominated regime. This misperception fostered the pursuit of "closet diplomacy" instead of the cultivation of widespread contacts in the bureaucracy and in Congress.[22] Specifically, many high Mexican officials sought to forge close, working relations with their U.S. counterparts with whom they could communicate regularly through personal contacts, telephone calls, or telex exchanges. Occasionally, these officials or their lieutenants emerged from the closet to lead delegations to the neighboring country to discuss, for example, salient provisions of a pending energy transaction. Unquestionably, such high-level diplomacy facilitated rapid decision making —particularly on the Mexican side—for a cabinet secretary or undersecretary had only to win approval of agreed-upon terms by a handful of individuals, the most important of whom was the president.

Still, disadvantages marred the success of closet diplomacy. Some important actors found themselves at the margin of, or excluded from, decision making. Limited contacts, typically by telephone, minimized each party's learning about conditions abroad, a situation tending to reinforce prevailing stereotypes. Too, those meetings that did occur frequently involved high stakes—with the possibility of a resounding, highly publicized success or an equally sensationalized failure. Also frustrating to Mexican devotees of high-ranking contacts was the rapid turnover in officials at senior levels in the U.S. executive branch where the revolving door swings rapidly. As one diplomat expressed it, "The investment in personal cultivation can rapidly evaporate leaving the 'closeteer' before a closed door with no handle."[23] Finally, any closet session eventually became known to the other potential participants within the U.S. bureaucracy. Jealous of their own turf, they demanded that future discussions

TABLE 8
Coincidence of U.N. General Assembly Votes Cast by the United States and Other Western Hemisphere Nations, 1975–1986
(in percent)

	1986 (200 votes)	1985 (201 votes)	1984 (192 votes)	1983 (179 votes)	1982 (202 votes)	1981 (160 votes)	1980 (107 votes)	1979 (123 votes)	1978 (132 votes)	1977 (98 votes)	1976 (92 votes)[1]	1975 (98 votes)[2]
Antigua and Barbuda	36.5	37.3	45.8	50.6	49.0	61.9	—	—	—	—	—	—
Argentina	29.2	30.1	26.8	30.2	31.4	39.7	43.9	45.5	57.6	53.6	46.7	53.1
Barbados	35.2	33.8	34.4	36.6	36.6	40.0	45.3	49.2	62.1	61.7	53.3	64.8
Bahamas	33.0	34.3	28.9	38.8	32.4	34.4	43.9	48.4	55.3	56.6	46.2	58.7
Belize	42.5	54.2	50.0	36.9	48.8	45.9	—	—	—	—	—	—
Bolivia	29.7	29.4	29.2	31.6	31.7	45.9	44.4	48.8	60.2	63.3	48.9	62.8
Brazil	33.2	31.8	30.5	32.7	34.7	36.6	39.7	45.5	59.5	56.1	47.8	54.1
Canada	66.7	66.7	68.0	69.0	65.6	71.9	81.8	81.7	84.8	83.7	70.7	76.0
Chile	45.0	41.0	38.3	41.1	41.6	47.8	50.0	52.8	62.5	62.2	55.4	66.8
Colombia	38.0	37.6	34.1	34.6	34.4	41.2	48.1	52.4	60.2	58.2	47.8	59.7
Costa Rica	47.5	44.5	35.2	38.8	38.4	43.1	46.3	48.4	59.5	64.3	54.3	71.9
Cuba	19.0	19.2	14.8	22.1	22.5	25.0	31.3	36.6	50.4	50.5	42.4	45.9
Dominican Republic	40.7	34.6	35.7	34.6	37.1	44.4	51.4	51.2	67.0	68.9	52.2	62.8
Dominica	60.2	54.5	59.1	55.0	56.4	62.8	66.4	66.7	67.8	—	—	—
Ecuador	35.5	34.3	32.8	33.0	32.2	34.1	43.0	48.0	59.8	59.2	47.8	54.6
El Salvador	49.5	39.1	38.3	37.2	39.4	40.9	44.9	53.7	65.9	64.3	52.7	63.3
Grenada	41.2	62.9	60.2	48.6	28.2	29.4	36.9	47.2	65.2	59.7	57.6	52.6
Guatemala	39.2	36.3	41.1	49.4	55.2	57.2	60.7	59.8	67.8	68.9	64.3	67.3

Guyana	22.0	24.9	20.3	26.5	27.0	32.5	40.7	44.7	56.8	55.1	43.5	52.0
Haiti	31.2	39.3	35.4	44.4	46.5	43.1	51.9	47.6	60.2	58.2	68.7	70.4
Honduras	42.5	40.8	37.8	41.1	43.8	44.7	45.8	54.9	65.2	63.3	64.3	68.4
Jamaica	34.2	35.3	31.5	33.8	31.7	44.1	38.3	45.9	55.7	54.1	44.0	50.0
Mexico	**28.7**	**25.1**	**19.3**	**27.7**	**27.0**	**31.2**	**41.6**	**47.2**	**59.8**	**60.7**	**44.0**	**51.5**
Nicaragua	20.2	21.6	17.2	24.9	24.3	28.4	35.5	44.3	69.3	67.3	61.4	78.1
Panama	32.7	32.1	32.0	32.7	29.5	33.4	42.5	48.0	59.5	59.7	46.7	54.6
Paraguay	40.0	51.2	45.8	50.8	50.2	57.8	55.6	58.1	65.2	70.9	65.9	67.9
Peru	28.5	29.1	27.6	33.0	31.7	36.2	44.9	47.2	57.6	55.6	45.1	50.5
St. Kitts	46.2	59.2	58.9	60.1	—	—	—	—	—	—	—	—
St. Lucia	42.0	42.3	47.4	45.8	49.3	40.6	48.6	63.8	—	—	—	—
St. Vincent	43.2	45.3	50.5	49.2	57.7	54.7	66.4	—	—	—	—	—
Suriname	29.2	29.4	27.6	30.4	32.2	34.1	43.9	49.6	61.7	64.3	50.5	74.4
Trinidad and Tobago	27.5	29.6	27.6	32.1	32.2	34.4	42.1	45.5	57.6	54.6	43.4	51.0
Uruguay	34.0	30.8	37.0	36.9	40.8	44.7	43.9	52.8	65.5	65.3	61.4	64.3
Venezuela	29.5	29.1	27.6	30.4	31.9	34.1	43.9	48.0	58.0	57.7	47.3	53.1
Average	36.9	37.8	36.7	38.8	38.6	42.3	47.2	51.1	62.0	61.4	52.7	58.9

Source: The U.S. Department of State supplied the data for this table. Rodney G. Tomlinson, professor of political science at the United States Naval Academy, performed the computer analysis that produced the percentages between U.S. votes and those of other nations of the Western Hemisphere. In making these calculations, a moderate approach was used—that is, countries whose representataives abstained or were absent for recorded votes were not penalized for disagreeing with the U.S. position. In cases of abstentions or absences, nations received "partial" credit for not directly opposing the United States as occurred when they cast "no" votes.

1. 91 votes for Guatemala, Haiti, Honduras, Panama, Paraguay, Peru, Suriname, and Trinidad and Tobago.
2. 86 votes for Suriname.

become part of the interagency process where debate, delay, and compromise—if not deadlock—flourished.

Comparison Between Canada and Mexico

The Canadians are as active as the Mexicans are passive in promoting their interests in Washington. Obviously, Canada enjoys several advantages over Mexico in its bilateral relations with the United States. Among these are a mutual enthusiasm for democratic institutions and capitalism, a cultural affinity, a shared language (except for the Quebecois), a history free of armed conflict since the War of 1812, fighting together in World War I, World War II, and Korea, and joint defense efforts in the North Atlantic Defense Command, and common membership in the North Atlantic Treaty Organization. Further, Canada's relative affluence—its 25 million people, less than one-third of Mexico's population, boast a per capita income of $12,940 (1984) compared to $2,200 below the Rio Grande—mitigates such problems as debt, drugs, and demographic factors, except when unlawful aliens attempt to enter the United States from a Canadian province. Knitting the bilateral ties even closer is the fact that four U.S. senators (Chafee, Leahy, Levin, and Lugar) are married to, or have close relatives who are Canadians. In all fairness, it should be noted that Mexican-Americans dominate the thirteen-member Congressional Hispanic Caucus.

Many problems bedeviling U.S.-Mexican affairs also affect those between Canada and its southern neighbor. Among those are trade, protectionism, investment, pollution, fishing, and U.S. cultural penetration. Like Mexico, Canada is affected acutely by Washington's actions or, to paraphrase former Prime Minister Trudeau, it is like a mouse and an elephant trying to sleep in the same bed. Traditionally, Ottawa has dealt with the elephant in the same understated manner now typifying Mexican diplomacy. As recently as the 1970s, Canadian embassy regulations prevented most Canadian officials from traveling the seventeen blocks from their Massachusetts Avenue chancery to Capitol Hill. When bilateral problems arose, a Canadian diplomat would register his country's viewpoint with the appropriate State Department desk and, possibly, with a White House official.

The emergence of protectionist pressures in response to a growing trade deficit helped persuade Canada, the United States' second largest commercial partner, to change its lobbying tactics. The new phase in Ottawa's Washington diplomacy began with the December 1981 arrival of Allan E. Gotlieb as Canada's ambassador. Rather than tiptoe around the capital and craft discreet notes to the State Department, the bespectacled Harvard Law School graduate and former Rhodes Scholar soon inserted himself into the power game. As he explained his role to a U.S. journalist: "Your constitution works on the basis of deal making. The administration can't govern by itself. The Congress can't govern by itself. You need a 'treaty' to govern. I get caught in that. The administration can't move the Congress, so I've got to get my hands dirty and try to move Congress myself."[24] And move Congress he did, thanks to an excellent flow of information from public opinion surveys and well-connected individuals whom he has befriended, his mastery of Washington's social life, his aggressively suave lobbying techniques, and the assistance of his wife Sondra, an iconoclastic novelist who contributed regularly to the *Washington Post*. "No ambassador," according to Senator Alan Simpson of Wyoming, the Republican whip, "understands the jungle of our politics as well."[25]

Eventually, sharp-fanged creatures within this jungle bloodied the energetic envoy, whose government hired Reagan confidant, Michael K. Deaver, to lobby in behalf of Canada on the highly sensitive acid rain issue. Soon Gotlieb found himself swept up in charges that Deaver had violated federal law by accepting the $105,000 assignment while still the White House deputy chief-of-staff. Gotlieb said that a remark by an unnamed Canadian official to public servant Deaver about Canada's being able to use a man of his talents was made in a "light-hearted, conversational" vein and was "hardly the stuff of negotiations or offers or proposals, of which there were none."[26] The ambassador suffered another setback when a distraught Mrs. Gotlieb slapped her social secretary before guests at a dinner party for Prime Minister Brian Mulroney.[27] Such tempests aside, the Canadians enjoy a superb reputation for exerting influence in Washington. Illustrative of the positive state of Canadian-U.S. relations was the signing, in January 1988, of a comprehensive trade accord that would eliminate all bilateral tariffs over a ten-year period, beginning January 1, 1989.

An insight into Canada's success vis-à-vis Mexico emerges from tables 9–12, which focus on the activities of agents who have registered with the Justice Department under the Foreign Agents Registration Act to advance the interests of private and public clients from the two countries. First, the number of agents for Canada tripled between 1978 and 1985 (the last year for which data was available), while the total for Mexico remained stable. Second, a notably higher percentage of Canada's agents were active compared to those registering in behalf of Mexican entities. Third, public and private Canadian firms expanded greatly the funds earmarked for political (Political) and quasi-political (Trade and Countervailing Duties) pursuits—namely, trying to influence officials in the legislative and executive branches—whereas Mexican entities continued to disburse only several percent of their monies on such important purposes. Fourth, Canadian clients consistently channeled 15 to 25 percent of their resources into attracting investment to their country, as well as fostering

TABLE 9

Pro-Canadian Lobbying Activities in the United States, 1978–1985

	No. of Registered Agents					No. of Clients		
	Active	(%)	Inactive	(%)	Total	Public	Private	Mixed
1985	78	88.6	10	11.4	88	35	89	0
1984	63	91.3	6	8.7	69	33	56	0
1983	53	91.4	5	8.6	58	26	37	0
1982	40	85.1	7	14.9	47	28	30	1
1981	39	84.8	7	15.2	46	29	24	1
1980	31	91.2	3	8.8	34	28	18	0
1979	30	96.8	1	3.2	31	28	15	0
1978	26	89.7	3	10.3	29	20	15	0

Source: Reports submitted to the Registration Unit, Internal Security Section, U.S. Department of Justice, Criminal Division, under the Foreign Agents Registration Act of 1938, as amended.

joint economic ventures between Canadian and U.S. firms; both activities enlarged a key North American constituency sensitive to the impact of Washington's policies on its northern neighbor. Meanwhile, between 1978 and 1985, Mexican clients expended but a pittance on these endeavors—a fact that may be related to the harsh treatment that Mexico suffered at the hands of Helms and others. Finally, funds spent by Canadian clients for tourist promotion declined steadily in recent years even as their Mexican counterparts—who for the first time in 1985 surpassed the Canadians in overall outlays—continued to concentrate more than 90 percent of their outlays on this economic sector.

Embassies form the centerpiece of a foreign country's efforts to radiate influence in Washington. The Canadian embassy has adroitly courted U.S. officials. In cooperation with thirteen consulates around the United States, the embassy has encouraged corporations with a stake in the outcome of a given issue to become involved in the decision-making process; it has also put businesses with a stake in the outcome of legislation in touch with politically well-connected legal firms. Mexico does have its largest embassy in the world in Washington. It boasts approximately 100 employees in the chancery, counselor section, attaché offices, and other dependencies—a figure that doubled between the mid-1970s and the mid-1980s. In addition, Mexico has eleven general consulates, to which are associated twenty smaller consulates, in the United States. Mario Rodríguez, the embassy's University of Texas–trained trade attaché, reported that in the early 1980s he was sent to the United States "to change . . . attitudes and communicate better" with Washington's power circle. To accomplish these goals, he doubled the size of Mexico's Trade Office in Washington, computerized his files, and endeavored to keep in touch with key congressional aides. This stepped-up diplomacy was done quietly. "It's our way of solving problems without creating others for ourselves," said

TABLE 10

Expenditures on Pro-Canadian Lobbying Activities in the United States, 1978–1985

(in dollars)

	Business Promotion[1]	(%)	Legal/ Financial[2]	(%)	Tourism/ Travel[3]	(%)	Trade and Countervailing Duties[4]	(%)	Political[5]	(%)	Miscellaneous[6]	(%)	Total
1985	1,880,618	15.7	1,197,430	10.0	2,232,617	18.7	2,432,449	20.3	3,033,246	25.4	1,183,315	9.9	11,959,676
1984	2,856,530	17.8	1,429,138	8.9	3,663,292	22.8	3,208,279	20.0	2,959,909	18.4	1,925,973	12.1	16,043,122
1983	2,203,247	15.5	1,376,097	9.7	2,684,507	18.9	4,115,242	29.0	1,512,867	10.7	2,287,359	16.2	14,179,314
1982	1,950,181	15.9	2,020,285	16.5	2,286,769	18.5	1,984,157	16.2	2,069,495	16.9	1,980,218	16.0	12,273,123
1981	1,766,720	25.1	367,611	5.2	1,997,775	28.3	1,533,410	21.7	360,495	5.1	1,025,379	14.6	7,051,390
1980	1,803,218	19.1	838,174	8.8	3,044,498	32.1	1,837,599	19.4	216,053	2.3	1,746,727	18.4	9,486,268
1979	1,392,914	15.0	444,771	4.8	5,147,137	55.6	1,240,306	13.4	476,872	5.1	559,456	6.1	9,261,456
1978	1,414,828	16.3	180,320	2.1	5,054,598	58.1	1,227,155	14.1	142,927	1.6	678,892	7.8	8,698,721

Source: Reports submitted to the Registration Unit, Internal Security Section, U.S. Department of Justice, Criminal Division, under the Foreign Agents Registration Act of 1938, as amended.

1. Includes attempts to encourage both U.S. investment in Canada and joint ventures between U.S. and Canadian firms.
2. Involves advising the client on legal questions and/or financial matters, such as the issuance of bonds.
3. Efforts to stimulate U.S. tourism in Canada.
4. Trade promotion and the representation of Canadian clients in countervailing duty suits before administrative and judicial agencies of the U.S. government.
5. Attempting to influence members of the legislative and executive branches on public policy issues, including bills pending before the U.S. Congress.
6. Promotion of cultural activities, collection of information, etc.

TABLE 11
Pro-Mexican Lobbying Activities in the United States, 1978–1985

	No. of Registered Agents					No. of Clients		
	Active	(%)	Inactive	(%)	Total	Public	Private	Mixed
1985	24	66.7	12	33.3	36	20	16	0
1984	16	41.0	23	59.0	39	12	14	0
1983	31	79.5	8	20.5	39	29	8	0
1982	34	81.0	8	19.0	42	30	9	0
1981	37	92.5	3	6.5	40	34	11	0
1980	39	86.7	6	13.3	45	33	13	0
1979	38	97.4	1	2.6	39	35	13	0
1978	33	89.2	4	10.8	36	38	11	0

Sources: Reports submitted to the Registration Unit, Internal Security Section, U.S. Department of Justice, Criminal Division, under the Foreign Agents Registration Act of 1938, as amended.

Rodríguez.[28] Still, he was only one man and found it impossible to cover all the political bases. After Mexico banned U.S. trucking services on the strength of an obscure 1955 presidential decree, the Teamsters Union and the American Trucking Association struck back by convincing Congress to pass the 1984 Motor Carrier Safety Act. Few Mexican truckers could meet the strict standards embodied in the legislation in order to operate on U.S. highways. Ultimately, the Mexicans forfeited the legislative contest because, at the insistence of the Ministry of Communications and Transportation, none of its representatives appeared at Senate hearings on the safety bill, which—in the absence of opposition—passed easily. "If you don't show up, the guy who does wins," stated a lobbyist employed by several Mexican clients. "That's the way the system works."[29]

Pemex's "Washington Beachhead"

Dismay over the embassy's low-keyed style prompted Pemex to broaden its presence in the United States. For some years, Pemex had operated offices in Houston and New York. The former, which opened in the early 1960s and expanded sharply under Díaz Serrano, concentrated on purchasing U.S.-produced equipment, supplies, services, and petrochemicals that the state company needed for its operations. The latter, which was established in 1943 and whose staff numbered forty-six people in early 1983, collected technical information on oil and financial markets, assisted in financial transactions with the major banks from which the monopoly had borrowed billions of dollars, and helped work out arrangements for sales to the monopoly's largest clients, many of whom are headquartered in New York. Invoices dispatched by the New York office to the international oil giants generated an average of $1 billion per

TABLE 12
Expenditures on Pro-Mexican Lobbying Activities in the United States, 1978–1985
(in dollars)

	Business Promotion[1]	(%)	Legal/ Financial[2]	(%)	Tourism/ Travel[3]	(%)	Trade and Countervailing Duties[4]	(%)	Political[5]	(%)	Miscellaneous[6]	(%)	Total
1985	—		107,615	.8	13,424,285	96.6	45,375	.3	306,368	2.3	10,729	0	13,894,372
1984	—		253,208	3.5	6,517,595	90.0	142,284	2.0	322,383	4.4	8,083	.1	7,243,553
1983	—		304,439	4.5	6,014,011	90.0	331,947	5.0	42,288	.5	2,210	0	6,694,894
1982	—		63,927	.6	9,723,164	98.8	23,671	.2	54,462	.4	0	0	9,845,223
1981	—		176,688	1.4	12,301,090	95.5	202,194	1.6	110,009	.8	90,000	.7	12,879,981
1980	—		34,804	.3	11,855,793	93.4	442,071	3.5	308,403	2.4	45,623	.4	12,686,694
1979	10,394	.1	11,133	.1	7,694,452	88.6	236,909	2.7	621,461	7.2	111,250	1.3	8,685,599
1978	—		17,607	.2	7,991,968	95.4	52,940	.6	216,602	2.6	100,619	1.2	8,379,735

Sources: Reports submitted to the Registration Unit, Internal Security Section, U.S. Department of Justice, Criminal Division, under the Foreign Agents Registration Act of 1938, as amended.

1. Includes attempts to encourage both U.S. investment in Mexico and joint ventures between U.S. and Mexican firms.
2. Involves advising the client on legal questions and/or financial matters, such as the issuance of bonds.
3. Efforts to stimulate U.S. tourism in Mexico.
4. Trade promotion and the representation of Mexican clients in countervailing duty suits before administrative and judicial agencies of the U.S. government.
5. Attempting to influence members of the legislative and executive branches on public policy issues, including bills pending before the U.S. Congress.
6. Promotion of cultural activities, collection of information, etc.

month. These resources, transferred immediately to the Bank of Mexico, constituted Mexico's principal source of foreign exchange.[30]

Although performing invaluable functions, neither the Houston nor New York office was in a position to keep abreast of political and economic developments, which arise not only from the U.S. government, but also from a host of international organizations, foreign embassies, and public affairs offices of large corporations and interest groups such as the American Petroleum Institute (API). For that reason, Beteta authorized the opening of a Washington office in April 1983. He took this initiative on the strong advice of Alfredo Gutiérrez Kirchner, a tax expert with a law degree from UNAM. Gutiérrez Kirchner was a longtime friend of the director-general under whom he had served in the Finance Ministry, an astute observer of the United States, the son-in-law of Antonio Ortiz Mena, finance secretary under López Mateos (1958–1964), and president of the Inter-American Development Bank since 1971. While working both as executive assistant to the president of the Inter-American Development Bank (1971–1976) and as minister counselor for trade and fiscal affairs at Mexico's embassy in the United States (1978–1983), Gutiérrez Kirchner became convinced that Mexico should expand its activities in Washington. Beteta named Gutiérrez Kirchner Pemex's general representative in New York and Washington.

After opening the Washington office, Gutiérrez Kirchner met with Foreign Secretary Sepúlveda to minimize any friction between the Pemex office, for which its newly appointed head anticipated an ambitious role, and the Mexican embassy. As mentioned above, SRE continued to seek a role as Regional Leader for Mexico even as the dominant economic leaders in the government, including Beteta, favored their country's performing as a Responsible Debtor. Ultimately, Sepúlveda agreed to an arrangement whereby Gutiérrez Kirchner, in addition to his Pemex assignment, would serve as the embassy's minister counselor for petroleum and energy affairs. After all, the embassy already boasted minister counselors for political affairs, economics, and trade, as well as specialists focusing on commercial matters, tourism, technology, the National Railways of Mexico, and the activities of the Nacional Financiera Development Bank. In a position paper justifying the Washington operation, Gutiérrez Kirchner argued that he could establish a Pemex "beachhead" for a modest amount.[31]

In fact, he set up the Washington operation on a shoestring, after trimming expenses in New York where the previous Pemex administration had approved the purchase of a $6 million mansion to house that office.[32] The combined 1984 budget for the New York and Washington offices was approximately half that for the New York office alone the previous year. Such savings enabled Pemex to open a field office in Madrid to complement those in Paris, Tokyo, and the

United States.[33] Also anticipated was the establishment of a London office, which—as of early 1988—had not occurred for lack of funds; retrenchment also forced closure of the Paris office.

Economies accomplished in New York facilitated the opening of the Washington office without the need to request additional funds from Pemex headquarters where Donanciano Támez Fuentes, the commercial subdirector, and others expressed reservations about a Washington operation. In addition to Gutiérrez Kirchner who spent three days each week in the U.S. capital, three professionals initially shared an incredible burden of responsibilities. One concentrated on U.S. economic policy and such salient financial matters as credits extended to Mexico by the IMF and the Inter-American Development Bank. A second followed energy developments in Washington, keeping in close touch with individuals in the departments of Energy and State— especially within the latter's Bureau of Economic Affairs. The Strategic Petroleum Reserve, to which Mexico sold 206.5 million barrels or 38.8 percent of the total volume acquired by late 1987, is of special interest. Moreover, he maintains contacts with the American Petroleum Institute, seventeen of whose sixty oil company members were Pemex clients. Indeed, senior Pemex executives attend the annual API meeting, which provides an opportunity to hold breakfast, luncheon, and dinner gatherings with, respectively, their small, medium, and large clients. The third professional specialized in trade questions, with emphasis on those linked to energy. Of most concern has been legislation embodying the natural resources subsidy concept. Such a measure would impose a countervailing duty on any imported good that benefited from a general subsidy such as, in Mexico's case, favorable prices for energy sold by Pemex or the Federal Energy Commission. Other interests include initiatives to (1) levy an oil import tax, (2) establish a Border Trade Zone, (3) convene a bilateral trade summit, (4) exempt from quotas the made-in-USA percentage of textiles or other finished goods imported from Mexico, (5) accelerate from six to nine months to twenty-one days the waiting period before the Secretary of Agriculture can restrict imports of perishable goods deemed disruptive of the U.S. market, and (6) limit the president's flexibility in permitting imports that technically violate U.S. trade regulations.

Although cutbacks in 1985 and 1986 reduced its full-time staff to one professional, Rafael G. Quijano, the Washington office of Petróleos Mexicanos continued to engage in manifold activities. It served as a listening post for the monopoly, while collecting and reporting to Mexico City information on energy and related developments in finance, commerce, and politics. For example, observers on Capitol Hill kept Beteta and his successor informed of the introduction of relevant legislation and changes in status of pending bills. Gathering and transmitting information laid the groundwork for attempts to influence pivotal decision makers. The Pemex office did this through direct contacts with bureaucrats, legislators, and congressional staff members. Further, it placed high Mexican officials conversant with energy issues in touch with those who shape U.S. policy.

A prime example of this took place when Beteta made a two-day visit to Washington in late March 1985. The director-general's busy agenda included two breakfast meetings with congressmen: the first, arranged by Rep. Robert Livingston (R-La.), attracted seven members of the House of Representatives; the second, sponsored by Sen. Lloyd Bentsen (D-Texas), enabled Beteta to exchange ideas with fifteen senators. Other activities involved meetings with Secretary of Energy John S. Herrington, ranking officials in the departments of State and Treasury, and William H. Drapper III, then president of the Export-Import Bank. Beteta also attended luncheons at the American Petroleum Institute and the U.S.-Mexico Chamber of Commerce where he delivered a speech on bilateral oil and trade relations, touching on U.S. protectionism, Mexican import substitution, and his nation's attempts to diversify exports. The Pemex head took time from meeting high officials and dealing with issues of broad significance to seek an interview with David Mulford, Assistant Secretary of Treasury for International Affairs. The purpose of this session was to resist the U.S. Customs Service's reclassification of Mexican exported low octane leaded naphtha from a motor fuel to a chemical mixture—a change that would have produced a fivefold tariff increase. Beteta's only misstep during the visit occurred when he conveyed de la Madrid's misgivings to the State Department about the attention that narcotics trafficking in Mexico was receiving in the United States in the aftermath of the murder of Enrique Camarena, an agent of the U.S. Drug Enforcement Administration (DEA). While recognizing that the visitor was only delivering a message, State Department officials believed his broaching the drug issue to be both gratuitous and irrelevant to his mission.

Gutiérrez Kirchner and his staff contacted all of Mexico's customers with U.S. operations concerning the natural resources subsidies bill. These clients ranged from multinational champions of free trade (Exxon, Shell, Mobil) to neutrals (Marathon, Amoco, Arco, Koch Industries) to arch protectionists (Texaco). In addition, Mexico kept in touch with Venezuela, Canada, and Saudi Arabia on the natural resources subsidies bill. Gutiérrez Kirchner and his counterparts from the other countries occasionally conferred on developments related to the legislation, even though conflicting interests militated against a coalition. For instance, while Pemex was concerned about petroleum, the Canadians keyed their efforts on timber, an item of little or no consequence to the Mexicans.

The subsidies bill was by no means the only object of Mexican interest.

Indeed, one of the most effective lobbying ventures involved tubular steel products. In 1983 U.S. Steel sought the imposition of countervailing duties ranging from 30 to 58 percent on Mexican steel imports. Pemex's Washington office discovered that U.S. Steel's most prosperous subsidiary was Marathon Oil, acquired in 1982. A longstanding Pemex client, Marathon was anxious to expand its purchases of Mexican crude. Because of the petroleum company's virtual autonomy within U.S. Steel's corporate structure, the oilmen had not consulted the steelmen about policy toward Mexico and vice versa. Hence, the Mexican government urged Marathon executives to become "our ambassadors to U.S. Steel." After discussions in Washington, New York, and Mexico City, the giant steel company softened its stance with respect to Mexico which, in fact, was a net importer of American steel products, with net purchases totaling 68,000 tons in the first half of 1983. The International Trade Administration did assess a countervailing duty of 5.65 percent ad valorem on Mexican oil-related tubular goods. Ultimately, however, Mexico signed a highly favorable, five-year volume restraint agreement with the U.S. government.

Pemex's Washington office also worked with Arent, Fox Kintner, Plotkin and Kahn, one of the U.S. firms that handles legal affairs for the monopoly. An example of the firm's activities was its role in collecting $1.3 million from Turkey in 1984 for a shipment of anhydrous ammonia made six years before.[34]

The Washington office distributed a monthly newsletter, *Pemex: Information Bulletin,* to keep government, business, press, and acadmic specialists informed of energy developments in Mexico, bilateral trade issues, and matters of general interest to the two nations. First published in October 1983, the newsletter became more ambitious in its coverage—with approximately two-thirds of the articles concentrating on Pemex and oil and the rest dealing with related trade and financial matters.

Information on prices published in the *Pemex: Information Bulletin* raised eyebrows in Pemex's Mexico City headquarters. Even though much of this information had appeared in Spanish in Mexico City, some Pemex executives apparently believe that it would remain within the country—a conclusion that ignored the large number of foreign journalists in the Mexican capital. On this and other issues pertaining to the Washington office, Gutiérrez Kirchner enjoyed Beteta's personal backing. Consequently, not only did the newsletter continue to appear each month, but its mailing list grew fourfold to embrace 2,000 names by 1987.

Gutiérrez Kirchner earned high marks from officials in the departments of State, Energy, and Treasury. They contrasted the drive, enthusiasm, and political acumen demonstrated by him and his colleagues with the "sleepiness" of the embassy, whose diplomats had qualms about Pemex's activism. With respect to the Pemex office's monitoring of the natural resources subsidies bill, the Pemex office, one Mexico specialist in the U.S. bureaucracy who asked to remain anonymous commented: "Those guys are incredible. They knew about the bill's being reported out of committee even before our own department's people assigned to keep tabs on the legislation became aware of the action." Other officials volunteered similar statements.

At least one group of congressional staff members do not share the bureaucrat's enthusiasm for the performance of either Pemex's Washington office, the embassy, or other spokesmen for Mexico who were attempting to shape the outcome of legislation. This conclusion sprang from the results of a survey, conducted in January 1986, of legislative assistants to members of the Ways and Means Committee.[35] This important committee considers tax legislation introduced into the House of Representatives. On its agenda for both 1985 and 1986 was the natural resources subsidies bill that was so important to Mexican interests. As indicated in table 13, advocates of Mexico's position wrote, phoned, or visited twenty-two of the thirty-six members of the entire committee, while they contacted nine of the representatives on the subcommittee handling the legislation.

Two contacts made the greatest impression: a small "informational breakfast," sponsored in September 1985 by the United States–Mexican Chamber of Commerce, that featured presentations by both Mexican officials and businessmen; and a letter that Rep. Esteban Torres (D-Calif.) sent to his colleagues. Except for these contacts, the staff members found the pro-Mexican lobbying to be unexceptional—with Mexican diplomats or lobbyists for the Mexican government making only six of the twenty specific contacts that could be recalled. When asked their opinion about the quality of these contacts, respondents rated only four of these contacts above average; that is, three were "somewhat effective" and one was "effective." None was considered "highly effective."

Undoubtedly, the frequent job changes and new assignments that occur

TABLE 13
**Contacts with Congressional Officers in Behalf of Mexico
on the Natural Resources Subsidies Bill, 1985–1986**

Has your office been contacted?	House Trade Subcommittee (14 members)	House Ways and Means Committee (36 members)
Yes	9	17
No	3	10
Unsure	0	4
No answer	2	5

Source: Telephone interviews conducted by Ms. Sue Sarnio, research assistant to George W. Grayson, in January 1986.

Public Relations Initiative

Increasingly, their friends in the United States urged Mexican officials to hire a public relations firm to devise a strategy for improving the country's standing and effectiveness in Washington. As one journalist expressed it, the image too often projected is that "Mexico is a country crawling with bandidos who prey on unsuspecting American tourists. Its smog-choked capital is a shambles in the aftermath of two devastating earthquakes. And drug traffickers, heavily armed and addicted to the good life, roam freely as long as they keep current on their payoffs to police."[38] In the aftermath of the scathing publicity beamed on its mid-1985 state and local elections, the Office of the Mexican Presidency entered into an agreement with the Hannaford Company, a "public relations and public affairs" firm with offices in Washington, New York, San Francisco, Sacramento, Madrid, and Taipei. "The number of negative stories grew in a way that worried us," said a Mexican official close to the president. "The U.S. press is the most hostile . . . and when you go personally to the United States to refute these charges, they tell you that it is too late. That is why it is important for us to be prepared." In other words, "When you go to Rome, it is important to do as the Romans do," he added.[39]

Peter Hannaford, the corporation's president, is a Californian who, since 1974, has served as a speechwriter, campaign advisor, and strategist for Ronald Reagan. He is author of *The Reagans: A Political Portrait,* an extremely flattering book about Reagan's quest for the White House.[40] Before launching his own firm, Hannaford was a partner with Reagan confidant, Michael K. Deaver, in Deaver and Hannaford, Inc., also a public relations agency. In selecting the Hannaford Company, which boasted Saudi Arabia and Taiwan among its foreign accounts, the Mexican presidency recognized the need to brighten its reputation north of the Rio Grande. Hannaford's activities in this regard increased after Mexico's Ministry of Commerce and Industrial Development terminated a three-month-long relationship with Deaver following the brouhaha over acid rain.[41] In return for its services, Mexico paid the Hannaford Company $30,000 per month plus expenses such as postage, air fare, and entertainment.[42]

What did Mexican taxpayers receive in return for this outlay? Above all, Hannaford sought to fashion a new image of Mexico. The Helms hearings stressed drug trafficking, corruption, and political turmoil. Mark E. Moran, the agency's general counsel and manager of the Mexican account, wanted North American opinion leaders to understand that, even with its problems, Mexico is a relatively open society that has enjoyed a half-century of political stability—a record unmatched in Latin America. "Continued talk of unrest could become a self-fulfilling prophecy," he told me.[43] The material distributed highlighted the nation's increasing devotion to free enterprise, its essentially healthy economy, its responsibility on the debt issue, and its growing flexibility toward foreign investors. "We see our role as interlocutors, assisting the Mexicans through the classic public relations methods," stated Moran. "I don't think the Mexican message has been getting across in the press. Maybe it's that people don't perceive the issues as being important or maybe it doesn't fit some preconceived notion. We don't think the Mexican story is being told in an objective fashion."[44] Hannaford sought to counter the flow of negative information about Mexico, stressing the basic points about four controversial questions: narcotics, corruption, immigration, and the Contadora process.

The gregarious Moran, who gained experience in international affairs as a staff member for the Senate Foreign Relations Committee, and his associates, Jared S. Cameron and Thomas Gorman, concentrated on the narcotics issue because of the attention lavished on it by the North American press, the DEA, and Senator Helms. The work on this issue was typical of their efforts in behalf of their client. First, they marshaled facts to demonstrate that Mexico had made a concerted effort to combat drug trafficking and that the United States' "demand pull" eclipsed its neighbor's "supply push" in explaining the accelerating northward flow of illegal substances. Information on the subject appeared in *Dateline: Mexico,* a "fact sheet" that pointed out, among other things, that the Mexican judicial police and armed forces suffered fifty-four deaths in 1984 and 1985 in their "war against narcotics" (compared to the deaths of three DEA agents); that as a percentage of national budget, Mexico's attorney general spent 50 percent more on drug enforcement than did the DEA; and that Mexico's Ministry of Defense earmarked more than $100 million annually to "fight drugs," while the U.S. Department of Defense spent only $15 million in 1984. The 1,500-person mailing list for *Dateline: Mexico* included all congressmen, influential members of the executive branch, editorial writers, wire service editors, television journalists, and selected members of the academic community. Hannaford hoped that the recipients would include the fact sheet in their resource file, to be consulted when next drafting a speech, writing a column, or preparing a commentary. *Dateline: Mexico* differed notably from the *Pemex: Information Bulletin.* The former appeared on an ad hoc basis as one or two pages of crisp, picture-free text, often organized as a series of bulleted points devoted to a single subject; in contrast, the latter appeared monthly, boasted several pictures on its eight pages, and sometimes employed turgid prose to examine a half-dozen or more subjects, some of which (such as trade legislation, Pemex's pricing mechanism, and foreign investment regulations) were quite abstruse.

Second, Hannaford complemented the fact sheet with news releases, distributed by mail and through the PR Wire Service conceived to help public relations specialists disseminate their message. Such materials kept the recipients abreast of developments related to Mexico's antidrug campaign—for example, the arrest of an individual suspected of murdering Camarena and his pilot, the confession of a close associate of Mexico's "drug kingpins," and the interception by Mexican authorities of a "cocaine caravan" containing 2,600 pounds of the illegal substance.[45] Third, upon learning of, say, plans by a senator to make a speech or a journalist to write an article on the drug question, the Hannaford executives attempted to provide Mexico's side of the story. Fourth, the public relations firm took the initiative by offering to set up interviews for prominent journalists with high-level Mexican officials knowledgeable about narcotics policy, the debt, energy, and other items of concern. For instance, Hannaford scheduled appointments for Dale van Atta (Jack Anderson's collaborator on the investigative "Washington Merry-Go-Round" column, syndicated daily to 1,000 newspapers) to meet with Silva Herzog, Beteta, Sepúlveda, and Salinas de Gortari.

Fifth, Moran singled out certain interest groups for special attention. High on his list were the United States' 8.8 million Chicanos who might be expected to develop the same empathy and understanding for Mexico that Jews exhibit toward Israel and Greek-Americans and Irish-Americans evince toward their respective ancestral lands. Several factors—lower educational levels, fewer resources, absence of a galvanizing leader, relative newness in the country of many of its members, and disagreements both within their own community and with other Hispanic groups—limited the power capability of Mexican-Americans. Yet, greater economic mobility, increased political participation, the reconciliation of internal divisions, and the acquisition of skills in coalition-building could bolster their position. Moreover, their numerical growth and geographic concentration should magnify their role as power brokers in pivotal Sun Belt states. Two mechanisms that reward bloc voting enhance this status: the one-member-per-district, winner-take-all mechanism for selecting members of the House of Representatives and the Electoral College for choosing presidents. Democratic and Republican politicians are keenly aware that California, Texas, New Mexico, Arizona, and Colorado embrace almost one-third of the electoral votes required to capture the presidency.

As a first act of Hannaford-inspired bridge-building to this constituency, the Mexican government sponsored a reception for the National Association of Hispanic Journalists (NAHJ), at its April 1986 convention held in Miami. Although the emphasis was social, the event provided Mexican officialdom an opportunity to convey its perspective on issues of interest to strategically placed Hispanics. The perceived success of the Miami gathering prompted the Mexicans to suggest that the NAHJ convene a future annual meeting in Mexico City.

Finally, Hannaford recognized the importance of educating Mexican leaders on the most effective way of achieving their objectives in Washington. To accomplish this goal and to keep open the lines of communication, Moran traveled to Mexico once a month. There he met with Manuel Alonso, the president's press coordinator, and other governmental leaders to analyze the actions of Washington policymakers and evaluate efforts to shape U.S. opinion with respect to Mexican policy on drug trafficking and other questions on the bilateral agenda.

The agency also choreographed the visits of high-level Mexicans to the United States, where it set up meetings with journalists from the print and electronic media. Moran helped arrange a February 1986 speech by de la Madrid to the California Newspaper Publishers Association. He also assisted in crafting the schedule for de la Madrid's August 1986 visit to the United States. In anticipation of what Hannaford termed a "summit meeting" between the two presidents, the agency dispatched an information kit to 2,000 U.S. opinion leaders. The "Dear Friend" letter that accompanied the material asked the provocative question: "Who is Miguel de la Madrid?" In response Hannaford Vice-President Cameron lauded the Mexican leader's technocratic background, extoled his accomplishments as Secretary of Planning and Budget, and applauded de la Madrid for having instituted "a sweeping austerity program" amid an economic crisis "similar to the Great Depression" in the United States. The praise turned to hyperbole near the end of the letter when Cameron stated: "On the political front, although burdened with vestiges of customs and habits dating back to the days of the Spanish viceroys, de la Madrid has worked toward 'moral renovation' of Mexican society. The government has recommitted itself to free and fair elections and is meeting once ingrained corruption with swift and effective punishment."[46] The chief item in the package was a thirty-two-page pamphlet, filled with color photographs, entitled, *Who is Miguel de la Madrid*[?].[47] Though attractive, this document, prepared by PRI's Secretariat of Information and Propaganda, was published during the 1982 presidential campaign and addressed few of the problems that confronted Mexico and the United States four years later.

Nevertheless, Moran reported positive results with respect to the public relations effort in behalf of the Mexican presidency. He cited as evidence of this success the flurry of editorials hostile to Helms's inquiry that appeared in leading newspapers (such as the *New York Times, Washington Post, Christian Science Monitor*) in the wake of the subcommittee deliberations. It is possible that sympathy for Mexico temporarily diminished the number of unfavorable articles about the fraudulent gubernatorial election held in Chihuahua in July 1986. Still, John J. Bailey, a Mexican specialist at Georgetown University, explained the diminished attention to the 1986 contests on the grounds that

North American reporters, who observed ubiquitous irregularities in the 1985 ballotting, concluded that it would be more of the same a year later and hence not a subject of major interest to their readers and viewers. Also encouraging to Moran was his perception that newspaper and television journalists had begun using in their reports the Hannaford-supplied data on, among other things, Mexico's war against illegal drugs.[48]

Yet law enforcement actions, presumably beyond the control of Mexico's chief executive, helped to undermine much of the work carried out by the Hannaford Company. When de la Madrid visited Washington in August 1986, the news broke that state police in Guadalajara had detained and tortured DEA agent Victor Cortez, Jr. Although the circumstances were murky, the incident stirred a firestorm of protest in the United States, including dozens of critical news and editorial reports. Eventually the criticism died down as media and mass attention shifted to the White House, the Middle East, and anti-Sandinista activity in Nicaragua because of the Iran-contra scandal. U.S. Ambassador Charles J. Pilliod, Jr.'s, low-profile approach (compared to his predecessor's more assertive style) removed an element of tension in bilateral relations as did Mexico's increasing readiness both to back market-oriented policies at home and to mute its criticism of Uncle Sam abroad. Additionally, the presidential selection process, completed in early October 1987, dominated the attention of most leading Mexican politicians.

In February 1987, Mexico moved its account from the Hannaford Company to Kindness and Moran, a northern Virginia firm engaged in legal work, public relations, and communications that Mark Moran, who left Hannaford in 1986, established with a former congressman. Late in the year, Moran continued to handle the Mexican account as he joined the Washington law firm of Shawn, Berger, Mann, and Moran. In what was largely a one-man operation, Moran scheduled loosely organized visits to Mexico for editors of regional newspapers, continued disseminating *Dateline: Mexico*, and began to initiate discreet contacts on Capitol Hill. As of late 1987, key U.S. officials involved with Mexico on a daily basis reported observing few positive results accruing from Moran's work. "Mexico really needs a high-powered firm to make its case effectively in this town [Washington]," one trade specialist said.

The Cortez affair indicated the enormous challenge facing a lobbying firm that sought to polish the image of a country whose police and other public servants engaged in practices considered repugnant by most North Americans. Assuming greater control could be exerted over offending Mexican authorities in the future—and this is a formidable assumption—the most effective champion of Mexico's interests in Washington would be a dynamic embassy complemented by efficient personnel in the Trade office (budget pressures forced the closure of the Pemex office on December 31, 1987). The effectiveness of these officials would be greatly improved if they could persuade business leaders with a stake in harmonious bilateral relations, including reasonable access of Mexican goods to the U.S. market, to contact U.S. policymakers—a task in which powerful Washington lobbyists and law firms can be extremely helpful. Failure to develop an effective presence in the U.S. capital militates against favorable laws and administrative acts, while opening the door both to greater protectionism and more Mexico bashing by individuals and groups who perceive that country to be a vulnerable target. Such actions will elicit recriminations from Mexico City and make it even more difficult for Mexican officials to promulgate the structural reforms essential to their nation's continuing to perform as a Responsible Debtor or adjusting to the status of Contentious Partner, a role discussed in chapter 8.

Mexico and the Organization of Petroleum Exporting Countries

Courtship

Mexico's massive oil discoveries of the 1970s took place during the Echeverría presidency. In view of the chief executive's penchant for activism in the international field, membership in OPEC seemed to be a logical step toward promoting his country as a leader of the region, if not of the entire developing world. In October 1974, Horacio Flores de la Peña, a self-described "leftist militant" who served as secretary of national patrimony, stated that his country would seek observer status in OPEC—with a view to joining the cartel as soon as possible. Meanwhile, Echeverría reportedly assured President Gerald Ford that Mexico would not affiliate with the thirteen-state entity which, to North Americans, seemed to combine the economic evils of cartels with the political dubiousness of "Third Worldism." The subsequent dismissal of Flores de la Peña temporarily laid the matter to rest.

At first, Francisco Javier Alejo, the new patrimony secretary and an Echeverría protégé, also turned thumbs down on membership. Yet, after a "confused melange of statements and counter-statements," Alejo announced in a May 5, 1976, press conference, "If OPEC countries think our entry would be useful, we will join immediately . . . [Although] we will not ask to join."[1] Echeverría nourished further speculation when, in Algiers during a forty-two-day tour of thirteen African, Asian, and Latin American nations, he said that as a small exporter Mexico did not then qualify for admission to OPEC, but that it hoped to become a member "as soon as it is possible." In the meantime, he endorsed a "substantial increase in the technological exchanges with OPEC members and total nationalization of oil resources to reduce the dependence of underdeveloped states on the major industrial nations and multinational companies."[2] For its part, Mexico would adhere to OPEC price increases. "We are not going to give our oil away to anybody," he added.[3] The ensuing ambiguity of Mexico's position preceded a firm rejection of OPEC membership by the López Portillo administration that assumed office on December 1, 1976.

The director-general of Pemex, the secretary of patrimony, and other prominent officials spurned the idea of affiliation. But the president himself delivered the most emphatic statement of his country's position in response to a reporter's question: "No sir," he said, "we are not going to enter OPEC. I see nothing at this time that necessitates joining the organization in the short, medium, or long term."[4] Subsequently he softened his comment by promising not to act like a "scab" toward an international commercial system that fixes prices.[5] Yet, López Portillo made clear that Mexico would abide strictly by its contracts to supply oil and shun any future cartel actions to halt exports for political reasons.[6]

The political attractiveness of OPEC was obvious. Membership would emphatically show Mexico's independence of the United States; indeed, joint action with thirteen other producers might even enhance Mexico's leverage in dealing with its powerful, energy-hungry northern neighbor. Moreover, the size and sophistication of Mexico's oil sector might have enabled it someday to play a key role within the cartel.

Doubtless, such factors weighed in the deliberations of Mexican policymakers. However, other considerations prevailed. First, Mexico has shown a traditional reluctance to affiliate with organizations that might compromise its freedom of action in international relations, especially on an issue as sensitive as oil policy. Mexico had not even joined the "Group of 77," a loosely organized and extremely diverse collection of developing countries, including all OPEC nations, that seek to advance their interests vis-à-vis industrialized countries. The autonomy that Mexico enjoyed over prices and trade in the petroleum field "would have been curtailed if we had joined OPEC," stated Díaz Serrano.[7]

Second, Mexico's state-owned industry was fundamentally different from that of OPEC countries in the 1970s. According to López Portillo, the cartel sprang to life because major producers, having given generous concessions to multinational firms, "wanted to fix oil prices themselves and not be at the mercy of these foreign companies." Even after the organization's formation, these companies continued to provide technical assistance, transportation, refining capacity, and market outlets. In contrast, Mexico has controlled its industry since 1938, and Pemex, a completely integrated firm, has become the largest corporation in the Third World. Moreover, the Mexican Petroleum Institute has more than 100 patents to prove its advanced status as a research and training center. Even Ahmed Zaki Yamani, Saudi Arabia's oil minister, emphasized how much his country could learn from Mexico in terms of technology in exploration, refining, and petrochemicals.[8] This ingratiating statement was undoubtedly true in light of the extremely low level of know-how possessed by OPEC countries.

Third, entry into OPEC would have complicated Mexico's relationship with the Third World which it hoped to lead. Industrialized nations could cushion the impact of higher energy charges by passing on these costs in the manufactured goods they sold abroad. Less developed countries (LDCs) enjoyed few such opportunities. The sharp rise in oil prices since October 1973 had ravaged their economies. Although OPEC countries pledged an additional $800 million in aid to developing states in mid-1979, the concurrent oil price increases meant that the cartel would return only one dollar for each additional ten dollars collected from less affluent customers.[9]

OPEC membership would have sparked tensions between Mexico and these hard-pressed oil importers. Outside the cartel, Mexico enjoyed the best of all worlds: the ability to avoid the negative aspects of affiliation while, thanks to lower transportation costs, charging more than most OPEC members for its oil. For instance, Díaz Serrano announced that as of April 1, 1979, Pemex's levy for light crude would be $15.36 per barrel compared to the cartel's reference price of $14.54.[10]

Fourth, OPEC membership could have affected Mexico's relations with Israel, to which—by the late 1970s—it was supplying 50,000 bpd of oil, a figure that had fallen to 40,000 bpd by 1987. Needless to say, Arab cartel members looked askance at such trading with their "enemy." Yet, apart from commercial considerations, its selling oil to Israel helped Mexico win favor with the United States, which—in the Camp David Accord—agreed to supply oil to Israel if it could not be obtained on the world market.

Fifth, from a purely legal point of view, Mexico did not satisfy—at least until the late 1970s—the prerequisite, set forth in Article 7(c) of OPEC's statutes, of "exporting oil in a substantial quantity." Although occasionally raised in discussions about Mexico's role with respect to the cartel, the "substantial quantity" concept is extremely nebulous and did not prevent the affiliation of such small producers as Qatar (1961), Ecuador (1973), and Gabon (1975).

Finally, OPEC is a complicated club, and affiliation would have required Mexico to take sides in intramural disputes. Remaining outside the organization enabled Mexico to avoid gratuitously alienating any cartel member or faction, while obviating reprisals from Washington. Under the U.S. Trade Act of 1974, the United States awarded preferential tariff treatment—the so-called General System of Preferences (GSP)—to selected exports of less developed countries. In retaliation against the 1973–1974 oil embargo, Congress excluded all OPEC nations from the GSP program. As a result, Venezuela and Ecuador, the two Latin American cartel participants, lost valuable commercial preferences even though both continued to ship oil to the United States during and after the 1973 Yom Kippur War. Congress did not repeal the "anti-OPEC" provision in the trade legislation until late 1979. Reportedly, at the time of Alejo's mid-1976 flirtation with OPEC, Treasury Secretary William Simon and Assistant Secretary Gerald Parsky warned Mexico that membership would result in the loss of GSP privileges. "I think the government sees the drawbacks involved," Parsky said.[11] Thus, Mexico, whose products enjoyed $458 million in GSP relief in 1978 alone, would have risked substantial losses by joining OPEC. In the words of Gregory F. Treverton, staff member of the National Security Council, membership for Mexico would have been "an economic catastrophe."[12]

Conflict

Mounting exports amid rising prices reinforced Mexico's decision to remain outside of OPEC, for there was little to gain—and much to lose—by compromising the country's freedom of action and complicating relations with Washington in a strong seller's market. Nevertheless, the López Portillo administration kept in touch with the cartel through discussions with Venezuela and, to a lesser degree, meetings with Yamani and other organization leaders. Mexico's correctness in relations with OPEC gave way to bitterness in late 1979 when the president presented his Global Energy Plan to the United Nations—a speech which reflected both his country's traditional championship of moral virtues and its emerging role as a regional, if not Third World, leader.

Behind the clichés, hyperbole, and vagueness lay the premise of López Portillo's discourse: namely, that the uneven distribution of oil and gas had widened the chasm between producing and consuming nations, giving rise to the threat of worldwide conflict. As a petroleum-producing nation, Mexico sympathized with nations struggling to revalue their raw materials. Yet, it was imperative, he added, "to dismantle a bogged-down system that works to the detriment of all . . . to cut this Gordian Knot."[13]

In a single paragraph, he proceeded to state that failure to design a rational international scheme would spark a "stupid holocaust," which would find humankind pointlessly suffering the "punishment of Sisyphus," thereby again loosing "the Horsemen of the Apocalypse"—this time, however, riding the unleashed energy of millions of horsepower.[14]

This apocalyptic danger, according to López Portillo, loomed largest in the next several decades, as the world underwent a transition from the petroleum age to the next era when solar, nuclear, and geothermal energy would be abundantly available.

The Mexican leader, who had earlier in his speech emphasized the importance of both the Golden Rule and the reconciliation of "conscience and national values with the interests of fertile and harmonious internationalism," proposed "a formula of teamwork, aimed not at imposition or intervention, but at harmonious participation that will integrate and amplify isolated efforts."[15]

What did he have in mind? Specifically, López Portillo suggested that just as the Bretton Woods Conference of July 1944 had spawned a comprehensive plan for handling postwar monetary and reconstruction matters through the World Bank, IMF, and GATT, the U.N. should give impetus to "an all-encompassing and balanced joint development strategy" in the energy field.[16]
Indispensable to any such plan were programs to:

- Guarantee the sovereignty of each nation over its own natural resources
- Provide financial and technical assistance to rationalize the exploration, production, distribution, consumption, and conservation of petroleum and other present-day energy sources
- Systematically develop potential traditional and nonconventional energy supplies
- Make it possible for all nations to draft their own energy plans to promote compatibility with the global plan
- Facilitate the production of capital goods and other energy-related manufactures in developing countries
- Resolve, in the short term, the problems of Third World oil importers, including access to supply, compensation for price increases, and "considerate treatment on the part of the exporting countries"
- Create a fund, with contributions from both developed consumer countries and producer-exporters, to help meet the emergency and long-range needs of petroleum-importing LDCs
- Assure easier access to, and transfer of, energy technology
- Establish an international energy institute[17]

López Portillo endorsed creation of a working group to "prepare the documents and pertinent specific proposals" related to Mexico's suggested plan. Constituting such a body would be representatives of oil-producing nations, industrialized countries, and developing petroleum-importing states. "The challenge is for all of us," he concluded, "because we are all part of the problem, and therefore, we are all part of the solution as well"—a phrase that was López Portillo's campaign slogan in 1976.[18]

A mixed reaction greeted the Mexican leader's ambitious scheme. Underdeveloped nations cheered it wildly; they had nothing to lose and everything to gain by ingratiating themselves with a surging oil exporter whose leader spoke the idiom of a new international economic order. Industrialized nations, especially those seeking Mexican oil, reacted politely, characterizing the plan with such positive but vacuous adjectives as "thoughtful," "imaginative," and "bold." Most enthusiastic of all was President Carter, who called the initiative "the most profound and beautiful I have ever read." Such beauty must have been evanescent because Carter failed to mention the program in his memoirs, although he devoted seven pages to the shah of Iran's adventures and misadventures with the López Portillo regime. This episode soured the U.S. president on his Mexican counterpart.[19]

Spokesmen for OPEC countries were as condemning as Carter was laudatory toward López Portillo's brain child. From conservative Kuwait to radical Libya, government officials decried the idea of discussions that focused on energy to the exclusion of such other vital issues as trade, aid, investment, technology transfers, etc. "We don't like [the plan]. It will die," was the brusque reaction of Kuwait's U.N. ambassador, who reiterated OPEC's opposition to "any separate treatment of energy as a theme." Libya favors the North-South dialogue on various matters, avowed its foreign minister, "but we will never permit a separate discussion of energy." Such a venture "is going to create an enormous quantity of problems and if Mexico is interested in this [international energy discussions], it should enter OPEC as a producer," he added.[20] The revulsion expressed by cartel members revealed less concern over Mexico as a competitor in production than fear that the important Latin American country, which boasted a tradition of social reform, would organize a new bloc of developing nations committed to a discussion of energy to the exclusion of other sensitive issues on the international economic agenda. Such a movement could have divided the ranks of the Third World and permitted industrialized countries to join in "OPEC-bashing" without exposing themselves on other economic and financial issues. At the very least, it would have embarrassed OPEC, which posed as a defender of developing states vis-à-vis industrial nations yet provided little aid to compensate for the huge trade surpluses registered with its putative allies. Venezuela's Luis Herrera Campíns expressed the cartel's collective pique when he said: "Some day Mexico may need the solidarity of the developing countries when it feels particularly vulnerable. This must never be forgotten."[21] Intense lobbying by OPEC nations helped sidetrack the initiative, which was to be given to the Group of 77 as a working paper, studied, and then—if approved—presented to the U.N. General Assembly's Second Committee (economics) as an official resolution whence it might reach the General Assembly.[22] The initiative, which died unceremoniously, served as a testament to López Portillo's enormous ego and pretentiousness as he attempted to thrust Mexico into an ambitious international role.

Rapprochement

During a period of ever higher prices impelled by rising demand, Mexico could afford to behave cavalierly toward OPEC. Even after the appearance of the

1981 glut, to which surging Pemex exports had contributed, Mexico maintained an export price for Isthmus ($32.50) between March 1982 and November 1983 that undercut that of Arabian Light ($34), the cartel's reference crude, as Mexico reached a new export peak of 1.5 million bpd. Evidence of this export growth came in 1982 when Mexico supplanted Saudi Arabia as the largest supplier to the U.S. market, where customers were anxious to buy the so-called nonreligious crude produced outside the politically volatile Mideast. Furthermore, Mexico sold substantial volumes at attractive prices to the U.S. Strategic Petroleum Reserve—"a move which enraged many OPEC members who see the stockpile as an instrument with which the U.S. tries to undermine their position."[23] In the words of Gilberto Escobedo, Pemex's subdirector for commerce, Mexico was "becoming international very fast . . . in every way"; that is, selling products, producing petrochemicals, and embarking upon joint ventures.[24] In May 1982, Mexico turned down an invitation to join OPEC as an observer. "We would have very little to gain and a lot to lose" by taking this step, Escobedo said.[25] After all, a severe economic crisis had increased Mexican dependence on the United States, and Mexico's acceptance of observer status with the cartel would have infuriated Washington.

Nonetheless, the continuing surfeit provoked a reappraisal of the country's relations with the cartel. Clearly, the stakes for international cooperation had risen for Mexico. Whereas oil and gas exports generated only a small fraction of its foreign exchange earnings in the mid-1970s, by 1982 they accounted for three-fourths of export revenue—a level that approached that of OPEC members with whose petroleum sectors Mexico exhibited increasing similarities. "We're walking on thin ice, but the ice could get a lot thinner," one official said, referring to OPEC's failure to agree on production quotas that would help sustain sagging oil prices. "The outlook is certainly bleak."[26]

This reevaluation of Mexico's oil policy attended the election and installation of de la Madrid in December 1982. Before his inauguration, a special eight-member "energy committee," one of several specialized task forces examining key policy areas, recognized inter alia the need to "avoid conflicts with the rest of the producing companies."[27] According to this report, it behooved Mexico to leave open the possibility of entering OPEC and abandon its unilateral policy of isolation that was incompatible with its status as an important world-class producer and exporter of petroleum. Even if it did not affiliate with the cartel, the country should embark upon a "search for new forms of dialogue, cooperation and negotiation with OPEC, with some of its most distinguished members and with non-OPEC exporting countries."[28] The document urged competitive pricing and expanded storage and handling facilities to boost exports, but pointed out the self-defeating consequences of such a policy if it sparked conflict with other oil producers. Thus, forging collaborative accords and consultative committees with key OPEC states was essential to bringing petroleum marketing into step with foreign policy. Such a recommendation was, at best, naive in light of cutthroat pricing policies within OPEC combined with the escalation of the Iran-Iraq War.

De la Madrid and Beteta took these proposals seriously in fashioning an international energy policy. Cooperation replaced competition with OPEC as Mexico explored ways to help stabilize a world petroleum market whose collapse would have dire consequences for its debt-ridden economy. After all, between 1979 and 1982 petroleum consumption by noncommunist nations had shrunk 12 percent to 46 million bpd. During this period, OPEC's share of the market had declined from 31 million to 21 million bpd, a minority share (47 percent) of the noncommunist market for the first time since 1962. Saudi Arabia's output dropped 35 percent in 1982, to 6.3 million bpd, its lowest average in 10 years. Prospects for the cartel appeared bleak in the absence of a new production quota. In March 1983, after months of haggling, the OPEC members reaffirmed a decision reached a year earlier to limit their combined output to 17.5 million bpd and to adhere to a lower benchmark price of $29 a barrel. Producers promised to refrain from both discounting prices and producing more crude than their assigned quotas.[29]

Although not a party to this accord, Mexico followed the negotiations carefully. From February to mid-March 1985, Pemex held off quoting prices to buyers, awaiting action by the cartel. High representatives of the Mexican government attempted to keep abreast of market developments in meetings held with counterparts from Venezuela, Algeria, Saudi Arabia, Nigeria, Norway, and Great Britain in advance of the London negotiations. During that conference, Mexican officials in the British capital kept in close touch with the OPEC ministers.

Moreover, in the aftermath of the agreement, Venezuela—in an unusual statement affecting a country passionately jealous of its sovereignty— announced that Mexico would bring its Isthmus price ($29 per barrel) into line with OPEC charges. Subsequently, Pemex reduced the price of Maya. Five days later, at the forty-fifth anniversary of the nationalization of Mexico's oil industry, Beteta stated that Pemex would maintain its production and export levels in cooperation with OPEC. In May he and Labastida issued a statement reiterating this policy alignment:

> In spite of [the] current situation of excess demand for Mexican crude, we will continue to hold to this export target. In today's delicately balanced market, any attempt by a large exporter to maximize foreign exchange revenues in the short run increases the risk of a similar action by other

exporters that also require additional foreign exchange. Eventually, such behavior could affect the current price level. This would imply a further reduction in foreign exchange income for oil exporters.[30]

The statement reflected two simultaneous objectives: to signal to other oil exporters Mexico's own efforts toward price stabilization and to reassure the country's creditors who feared further decline in export income.[31] The de la Madrid government decided that Mexico's long-term goals would best be served through short-term sacrifices and continued to observe production ceilings. As one observer noted, "Mexico, another non-OPEC producer that has never paid much attention to the cartel, has done a total about-face. Although saddled with a foreign debt of $85 billion, Mexico is limiting its oil sales to 1.5 million bpd to support the OPEC price level."[32] This praise overlooked the fact that except on a surge basis, when exports could be boosted by 200,000 to 300,000 bpd, the 1.5 million bpd level was as much as Mexico could sustain in the absence of major new investments.

Mexico's rapprochement with the cartel continued as it became the first country to send informal observers to OPEC meetings. In July 1983 initial consultations focused on issues of technical cooperation, leading to subsequent discussions of pricing and production policies appropriate to shoring up the market.[33] Mexico carefully established an array of ties with OPEC producers and observed the organization's price and production guidelines. Its new role as a "kind of informal member of the cartel"[34] was a far cry from the confrontation evident during López Portillo's administration. The July meeting led to a price increase in Maya crude, comparable to Venezuela's charge for a similar variety. Having learned a painful lesson in 1981, Mexico began tailoring its policy to OPEC in an atmosphere of cooperation and solidarity. The price increase served political and commercial objectives at a time when the demand for Maya was particularly strong because large investments in refineries had enhanced the attractiveness of relatively cheaper heavy crude, the greatest volume of which—the refiners knew—lay outside OPEC countries.[35] In fact, in December 1983 Pemex reported that it had turned down orders as high as 400,000 bpd to avoid exceeding its 1.5 million-bpd export quota.[36]

Throughout 1983 and 1984, Mexican representatives met with officials in producing countries to exchange information, explore proposals to promote market stability, and encourage communications between OPEC and non-OPEC countries. Labastida took the lead in an effort that focused on stimulating cooperation between Canada, North Sea exporters, and the cartel. In mid-1983 in Caracas, the Mexican energy secretary emphasized that Mexican membership in OPEC was far less important than a display of harmony with both cartel and noncartel exporting nations.[37] What Labastida failed to say was that, in view of U.S. support on debt matters, including SPR purchases from Pemex, Mexico's joining OPEC was out of the question.

In August Labastida met with his counterparts from Venezuela, Ecuador, and Trinidad and Tobago to explore the establishment of an Organization of Latin American Petroleum Exporting Countries (OLAPEC). Once in place, such an OLAPEC would play the same regional role as the Organization of Arab Petroleum Exporting Countries (OAPEC), stated an official from Venezuela, the country that had spearheaded the initiative. Still, he neglected to point out that OAPEC had played no role at all since imposing an embargo on the United States and the Netherlands in the wake of the October 1973 Mideast War. Additionally, it was argued that the new organization would bring Mexico closer than ever before to the deliberations of OPEC by linking it to two cartel members.[38] Three months later the energy ministers convened in Cancún. At the end of a two-day session, the leaders called for talks with consuming countries and producers from other regions to stabilize the international market. While a Mexican spokesman insisted that conditions augured well for an international conference, the November 1983 proposal proved no more successful than had López Portillo's Global Energy Plan four years earlier.[39] Meanwhile, OLAPEC remained only a gleam in the eye of its Venezuelan advocates.

In early 1984, six influential OPEC ministers met informally with their counterparts from Mexico and Egypt. They concurred in the need for a stricter ceiling to defend the market price for crude oil. After the meeting, Yamani, accompanied by Mexican and Egyptian representatives, led a small delegation to Lagos in hopes of persuading the Nigerians to rescind the $2 per barrel price cut they had just implemented. The spirit of solidarity prompted Youseff M. Ibrahim, the *Wall Street Journal*'s petroleum analyst, to comment: "In the new order that is being shaped, oil producers, whether inside OPEC or outside the cartel, are banding together to uphold oil prices. The task, which used to be OPEC's alone in the 1970's, is now shared by a vast array of non-OPEC producers with differing ideological convictions such as Britain, Mexico, the Soviet Union and Egypt."[40]

Later in the year participants in the twelfth annual OPEC Workshop for Journalists, held in Mexico City, lavished praise on their host as a prime mover in this new order. OPEC's president, Kamel Hassan Maghur, stated: "Mexico is a partner and not a competitor of the Organization of Petroleum Exporting Countries. . . . Mexico has adhered to a policy that strengthened OPEC decisions, enhanced South-South cooperation, and enabled others to learn from the Mexican experience."[41] In a similar vein, Mohamed Charara, Saudi Arabia's ambassador to Mexico, observed that Mexico "is cooperating" with OPEC on production and prices "much more than the North Sea producers," and that without such cooperation with OPEC pricing policies, current pricing and production stability would not exist.[42]

Still, it was the objective supply-and-demand situation, not these flattering words, that encouraged Mexico to express solidarity with the cartel in hopes of strengthening a market debilitated by most OPEC members' readiness to exceed their assigned quotas by offering discounts, special trade arrangements, and other inducements. Yamani flew to Mexico City in September 1984 to seek Mexico's cooperation with an action, taken in late October in Geneva, to pare the cartel's production ceiling from 17.5 million bpd to 16 million bpd to halt the slide in prices. Mexico dutifully responded that it would, effective November 1, trim its own export level by 100,000 bpd. Labastida and Beteta described the move as a "substantial sacrifice" that would deprive their country of $81 million in foreign exchange. Still, they added, "instability affects all oil exporters, not only OPEC members. That is why we all must contribute to the market's orderly development and share the burden this effort implies."[43]

Such self-serving proclamations aside, the action appeared more political than economic as Mexico made virtue out of necessity. Pemex's exports for the first ten months of 1984 had averaged 1.543 million bpd; hence, some year-end reduction was crucial to Mexico's staying within its self-imposed 1.5 million bpd export ceiling. Too, Pemex began encountering market resistance in August, and a ballyhooed production cut in concert with OPEC's market stabilization plan appeared preferable to lowering prices.[44] Finally, bad weather often causes a reduction in Mexican production for a day or two every six weeks between September and March as tropical storms pummel Gulf ports, including the terminal at Pajaritos from which Pemex dispatches the majority of its crude exports. For example, in January 1985 such storms delayed the lifting of some 3 million barrels of crude as tankers formed a line along the five-mile channel leading to the port of Pajaritos.[45] Still, some market analysts contended that Mexico decreased production by only half the proposed figure, although Pemex reported that sales were 1.438 million bpd in November and 1.440 in December 1984.[46]

Mexico Decries Cheating

By selling only under contract, spurning discount prices, and adhering to its export target, Mexico earned a reputation as the "best member of OPEC," even though it remained outside the strife-ridden organization. Such model behavior stood in contrast to practices followed by a majority of cartel members who flouted their October 1984 production quotas and price structure. For example, Algeria, Iran, Libya, and Saudi Arabia exchanged oil for industrial goods and arms in barter deals that disguised huge price discounts. To attract customers saddled with rising insurance premiums as a result of Iraqi air attacks on the Kharg Island terminal, Iran offered discounts of $3 to $4 a barrel on its oil. This was in addition to the $1 discount already allowed by OPEC.[47] By mid-1985 OPEC experts estimated that 80 percent of the cartel's total exports of some 14.5 million bpd were sold at below the organization's so-called official prices which averaged $27 per barrel.[48] Ecuador, Gabon, and other member countries sold to traders in auctions that ignored official prices. Or they gave petroleum firms that produced oil in their territories generous cost allowances, which amounted to another form of discount.

In making common cause with OPEC, Mexico soon discovered that it was being made a fool of by the cartel. "The Mexican government has realized that they are the only ones playing by the book," stated one official in Mexico City. "Pemex is tired of seeing OPEC members violating their own accords time after time. If OPEC countries can't sell at official prices, they have any number of ways to get around it . . . but not Mexico."[49] In a communiqué issued on December 22, 1984, COCEP bluntly warned that Mexico had had enough of the cartel's lack of discipline. Although pledging to maintain the pricing and production policy established in November, it deplored "widespread discounting and destabilizing commercial practices of OPEC which ran contrary to oil market stabilization policies." For that reason: "OPEC must restore discipline among its ranks or Mexico will take unilateral steps to defend its interests."[50]

Doubtless, Mexico hoped that its statement would chasten OPEC members so as to obviate the need for further threats. Mexico's importance to OPEC seemed to increase when Britain and Norway moved to spot market pricing early in 1985. "Mexico's threatened break would be an even bigger blow to OPEC, Mexico exports more oil than Britain and has been one of the cartel's staunchest supporters."[51]

Mexico attempted to bring Britain and Norway into OPEC's camp. Before an OPEC emergency meeting in late January 1985, Labastida and Navarrete met with oil officials from the two European producers in hopes of restoring their confidence in OPEC's efforts.[52] The venture proved fruitless; however, the Mexican delegation traveled on to OPEC nations—Saudi Arabia, Kuwait, Algeria, Indonesia, and Venezuela—to reiterate the warnings contained in COCEP's December communiqué.

One Mexican official said the outcome of the Geneva meeting would be "crucial" in determining his nation's attitude toward OPEC.[53] The meeting produced no real progress; only nine of thirteen members agreed to cut oil prices of Arabian light by $1, to $28 per barrel, and no anticheating program materialized. A system devised the previous December to monitor cheaters who exceeded their national oil production quotas or sold below a floor price turned out to be an exercise in futility. The logistical problems in keeping track of oil shipped from more than fifty ports were staggering. This is not to mention the challenge of dealing with thirteen nations—many of which thrive on secrecy—bitterly engaged in mutual accusations of skullduggery. It was these very states, including the cheaters, that had to provide information on pricing, production, and quality of crude if supervision was to be successful.

The response of an increasingly disillusioned Mexico soon followed. In an action described as "autonomous" and "in the national self-interest," Labastida and Beteta announced a reduction of $1.25 per barrel for Isthmus, bringing the price to $27.75 and costing Mexico an estimated $300 million per year in lost revenue. The reduction exceeded OPEC's, but Mexico faced the loss of U.S. customers attracted by low spot market prices. In addition, market resistance was building in Europe—due to the end of the coal strike in Great Britain—and in Japan. Conditions necessitated the price cut, for as one observer noted: "Faced with the threat of losing some long-term customers, Mexico must stay competitive. Otherwise it may slip on the road of economic recovery and stall in payments on its foreign debt."[54] A few days earlier, the discord and lack of discipline within OPEC also led Egypt—which like Mexico had frequently acted in concert with the cartel—to cut the price of its light Suez crude. News of the Mexican and Egyptian moves arrested the gradual rise in spot market prices.[55]

On February 5, 1985, SEMIP announced that the Mexican oil export platform would return to the 1.5 million-bpd level, with extra shipments in February to compensate for January's shortfall. This move exposed the rift between Mexico and the unruly cartel members. But officials also emphasized their constant interest in stabilizing the oil market. In the words of Labastida and Beteta: "The signal to the market is clear. Mexico will continue to maintain dialogue and cooperate towards market stability. It will also continue to take into consideration its own national interests as well as those of the oil-producing and exporting countries and the consumers themselves."[56]

Market analysts viewed the communiqué and subsequent actions as a firm warning to OPEC. A break was almost certain if the situation worsened.

> Mexico believes that discipline and coordination is the only way to safeguard oil prices and has found these qualities lacking in OPEC members. Mexico knows it has a lot of clout in world markets, but it expects OPEC to play it straight and stop undercutting the market with covert discounts and incentives. If they do not, Mexico can afford to boost production or drop prices further.[57]

The *Wall Street Journal* estimated the possible increased production at 250,000 bpd, an amount that would put even more pressure on falling world prices.

Even more worrisome to OPEC than these actions was Mexico's January 1985 unveiling of a revised five-year plan that called for a 3 percent annual growth in production—with increase in exploration, drilling, and refining. The program anticipated that output would reach 3.3 million bpd by 1990, with exports climbing to 1.8 million bpd—20 percent more than the 1984 average.[58] However, this announcement proved a Potemkin promise, for Mexico lacked the resources necessary to expand its industry. In addition, while revulsed by OPEC's behavior, Mexico continued to do everything possible to sustain oil prices. A precipitous drop would be, in de la Madrid's words, a "sign of international madness."[59] To avoid falling into the abyss of madness, Labastida and Sepúlveda urged Mexico to keep its prices in line with those of OPEC—advice cheered by the Mexican left, which believed their country should act as the cartel's fourteenth member.[60] As discussed in chapter 3, this policy turned out to be catastrophic, for Pemex lost customers to exporters offering cheaper prices. In June 1985 Mexico lost $515 million as sales abroad fell below 800,000 bpd—a decline that emphasized the need for market-sensitive pricing.

This shock precipitated a year-long experiment in independent pricing as Pemex sought to regain its market share in the face of expanded production by Saudi Arabia, Kuwait, and other Gulf states. The Saudis seemed determined to convince Great Britain, Norway, the Soviet Union, and other key non-OPEC producers that failure to cooperate with the cartel would exacerbate the surplus and drive prices to a ruinous level. By its action, Mexico became the last major non-OPEC exporter to abandon formally the cartel's supposed efforts to prop up prices. As one energy official expressed it: "There is a strong current of feeling in Mexico that it's time to look after our own interests."[61] Disenchantment with OPEC in general did not prevent an orchestrated price cut by Mexico and Venezuela in early 1986, when their presidents—meeting in Cancún—agreed to reductions of as much as $4 per barrel in an attempt to remain competitive, especially in the U.S. market where both sell large volumes of heavy crude. The two chief executives promised even closer cooperation on oil matters in the future, to be facilitated by an Oil Cooperation Board. This joint entity would enable Mexico and Venezuela to share information on markets, pricing, and production in order to "arrive at decisions that protect the interests of both countries."[62]

When carefully reasoned prose failed to convince Saudi Arabia of the self-defeating nature of its strategy, Mana Saeed Al-Otaiba, oil minister of the United Arab Emirates and author of fourteen books of poetry, composed the following verse, entitled "A Free Invitation to the Oil Banquet":

> Oh you who have built for OPEC a mighty place,
> Do you now seek to demolish its edifice?
> Is this a brave action or a streak of madness

Which people of intelligence could not contain?
Will you now turn your oil into a sword of war,
Hoping that its own sharp edge will defend it?
No! Do not use the sword of oil to hit against
Swords of others, lest you blunt your own instrument.[63]

However, it was not the muse but Iranian politicians who precipitated still another OPEC production ceiling. This one, which became effective September 1, 1986, imposed a 16.7 million bpd limit on collective production. The agreement was possible because of the replacement of Saudi Oil Minister Yamani, a staunch advocate of the "price-war" strategy, by Hisham Nazer who pursued a moderate course at the instance of King Fahd. A skeptical Mexico immediately expressed a willingness to cooperate with the cartel by decreasing its own exports 10 percent below the 1.5 million bpd target. Again, COCEP was making virtue out of necessity inasmuch as Pemex exports, which totaled 1.4 million bpd in July 1986, averaged only 1.3 million bpd in August and September.

In December 1986, OPEC reiterated its determination to adhere to quotas in order to lift prices. The cartel also decided to return, on February 1, 1987, to a fixed price for oil—namely, $18 per barrel, the weighted average of seven crudes, including Mexico's Isthmus. OPEC placed Isthmus in its pricing basket without consulting Mexico and at an unrealistically high price—$18.07 per barrel compared to $17.52 for the more desirable Arabian Light variety. Although irate at this transparent, Venezuelan-inspired move to influence the charge for Isthmus and to induce Mexico to discard its market-sensitive formula for official pricing, Energy Secretary Alfredo del Mazo announced a further reduction of 30,000 bpd, to 1.320 million bpd, in the country's self-imposed ceiling. In the face of its lower export quota, Pemex had to turn down some 500,000 bpd in purchase requests in February. Nonetheless, Mexico believed it imperative to follow a strategy that would enhance the market stability that OPEC had attained. Vindicating this decision was the fact that stronger demand elevated the price of Isthmus to above $18 per barrel in mid-1987, as exports averaged 1.337 million bpd for the January-June period. Still, del Mazo warned that, in the absence of concrete steps by OPEC members to uphold the new quota, Mexico might assume a larger leadership position among noncartel producers in order to facilitate "negotiations and presentations to OPEC members."[64] He failed to specify the probable content of such exchanges.

Mexico's complicated relationship with OPEC contributed to the country's pragmatism and role definition as a Responsible Debtor. Between the mid-1970s and mid-1980s, Mexican relations with the cartel took many forms: courtship and near-membership, conflict, cooperation, and correctness. Ultimately, Mexico realized that market forces played havoc with any cartel unless the dominant producer, in this case Saudi Arabia, adjusted its output to maintain prices. In addition, stentorian pronouncements about "maintaining unity," "preserving a community of interests," and "following common objectives" paled before the quest by each of the thirteen states to meet its own domestic financial needs. Belatedly, Mexico learned that its influence on the international energy stage, though greater vis-à-vis OPEC during a period of falling prices, was far less than López Portillo had imagined. Slavishly adhering to official prices and trying to play by rules that others were breaking in a world awash with oil cost Mexico billions of dollars. The expensive lesson conveyed was that national interests dictated selective cooperation with what is, after all, an uneasy grouping of diverse countries. Also apprehended was the importance of basing prices on supply-and-demand forces—a vital consideration for a Responsible Debtor, especially in view of overproduction and persistent cheating by key OPEC members in early 1988.

Mexico and the San José Accord

The fourfold increase in oil prices imposed by OPEC in late 1973 and early 1974 sent economic shock waves through the countries of the Caribbean basin. Not only did energy and other imports become more expensive, but an increasingly sluggish demand in recession-afflicted industrialized states retarded the growth of export earnings. Internationally, Caribbean area nations suffered gyrating changes in their terms of trade, as well as balance-of-payments deficits and mounting foreign indebtedness to public and private financial institutions. Domestically, they endured budget shortfalls, escalating prices, rising unemployment, and pressure—notably from the IMF and world bankers—to adopt drastic austerity plans.[1] Where could the hard-pressed countries of the region seek assistance that would not carry with it burdensome obligations that might intensify social problems? Was there a nation in the region which they could turn to for aid?

Not surprisingly, Venezuela appeared at the top of the list of potential benefactors. After all, the Christian Democratic President Rafael Caldera had inaugurated, with the help of his Trinidadian-born foreign minister Arístides Calvani, a new Caribbean policy rooted in geopolitical concerns. Specifically, the Venezuelan leaders wished to ensure safe passage of their country's petroleum shipments, promote political stability in poor and backward island states, and develop markets for such exports as processed foods, petrochemicals, textiles, and light manufactures.[2] In November 1971 Calvani convened a consultative meeting in Caracas of foreign ministers of Caribbean states; two subsequent sessions focused on regional transportation concerns; and, in April 1973, Venezuela became the first non–English-speaking member of the Caribbean Development Bank.[3]

As a leading OPEC participant, Venezuela benefited handsomely from the surge of energy prices. Thus, the Democratic Action government of Carlos Andrés Pérez, which won the national elections in 1973, greeted the request for assistance from its neighbors by creating a cash-loan plan to offset the rise in oil costs incurred by Panama and the five Central American countries.[4] Under the "First Program of Financial Cooperation" announced at Puerto Ordaz in December 1974, Venezuela agreed to allow beneficiary importers to keep in their central banks all monies above $6 per barrel paid for oil, for which the international charge was then $12. To finance balance-of-payments deficits, Venezuela would loan these retained monies to the buyers for six years; however, if a participating nation proposed suitable development projects, cosponsored by an international financial institution, it could borrow the funds for up to twenty-five years, with a six-year grace period, at soft interest rates equal to those levied by the Inter-American Development Bank in its ordinary capital operations, approximately 8.5 percent. The Venezuelan Investment Fund (FIV), its income generated from taxes on oil and gas sales, was given responsibility for approving projects under the Puerto Ordaz accord. This agreement, later expanded to include Jamaica and the Dominican Republic, covered a volume of oil equivalent to five-sixths of all imports in the base year (1974), then a gradually decreasing percentage of purchases until it expired on December 31, 1980.

Sixty-two percent of the $678 million committed and wholly disbursed under the program by January 1982 had been converted into long-term loans for projects in the areas of energy, water, agriculture, transportation, and industry. The allocation of resources was as follows: Costa Rica, 12 percent; El Salvador, 15 percent; Guatemala, 19 percent; Nicaragua, 11 percent; Jamaica, 8 percent; Dominican Republic, 6 percent; Honduras, 13 percent; and Panama, 16 percent.[5] By mid-1984 the disbursement level had reached 90 percent.

Originally, the Pérez regime viewed the aid scheme as a transitional measure to enable countries, heavily dependent on Venezuelan supplies, to adjust to higher energy charges. Yet, the doubling of oil prices in 1979 and 1980 because of the Iranian revolution and the subsequent Iran-Iraq War led Caracas to cast about for partners in a new venture that would help regional economies ravaged by the sharp increase in oil prices and permit Venezuela to share what was rapidly becoming a heavy burden in terms of both volume of crude supplied and financial stress.

In a parallel move, Venezuela attempted to convince the entire OPEC group to establish an aid program for all poor nations. When the cartel, whose Arab members prefer to assist their religious and ideological brethren, rejected the idea of its own oil facility, Mexico emerged as the prime candidate for inclusion in a regional undertaking. Mexico's production had climbed from 209,855 bpd in 1974 to 536,926 bpd in 1979, while its announced proven reserves had shot up more than eighteenfold to 45.8 billion barrels in the same period.[6]

Initially, pleas from governments of the region to furnish discounts or special arrangements fell on deaf ears in Mexico City. Patrimony Secretary José Andrés Oteyza stated: "Although they are needy, priority in selling them our oil will be determined by the terms of international trade rather than by any other consideration."[7] At least four factors prompted Mexico to reverse its "strictly business" stance.

First, as previously discussed, in September 1979 López Portillo launched his Global Energy Plan in a U.N. speech. Among other things, this proposal called for cooperation between producing and consuming nations, and the establishment of "a short-term system to be put into effect immediately, for resolving the problems of developing countries that import petroleum." If implemented, this system "would guarantee supply and the honoring of contracts, stop speculation, provide for compensation for price increases, and even ensure considerate treatment on the part of the exporting countries."[8] Attacking the "social and economic causes of problems" that afflicted the Caribbean basin could be a first step toward a more ambitious international energy program.[9] It would also demonstrate Mexico's commitment to aiding oil-importing nations of the Third World.

Second, on January 24, 1980, the Mexican chief executive spent nine hours in Managua where he condemned the "satanic ambition of imperial interests" and suggested that the Sandinista revolution—like the Mexican and Cuban ones before it—offered a promising path for Latin American nations anxious to escape the problems besetting the hemisphere. He offered assistance to the country's fishing and communications industries and pledged that Petróleos Mexicanos would supply 7,000 bpd of crude—one-half of the nation's consumption and an amount termed "indispensable" for the regime's survival.[10]

And third, just two weeks later López Portillo welcomed the Jamaican prime minister, Michael Manley, to Mexico City. The leaders discussed regional issues, stressed their support for ideological pluralism and self-determination, and announced increased economic cooperation. López Portillo agreed that Mexico would provide 10,000 of Jamaica's 27,000-bpd oil requirements in exchange for 420,000 tons of bauxite each year.[11]

Finally, the Costa Rican president, Rodrigo Carazo Odio, joined the Venezuelan government in urging Mexican involvement in an areawide assistance venture. Carazo's interest in such a program was sharpened when Shell of Curaçao, which had been supplying his country, canceled its contract in 1979, forcing Costa Rica to purchase crude on the Rotterdam spot market. Carazo's lobbying bore fruit on August 3, 1980, when López Portillo and Luis Herrera Campíns, a Christian Democrat who had succeeded Carlos Andrés Pérez as president of Venezuela, agreed to the Economic Cooperation Program for Central American Countries.

Structure

This program, commonly known as the San José Accord, represented the first collaborative aid effort between an OPEC and a non-OPEC country.[12] Under its terms, Mexico and Venezuela each pledged eventually to ship up to 80,000 bpd of crude oil, or "reconstituted" crude ("recon"),[13] on concessionary terms to nine nations of the region—specifically, those covered by the Puerto Ordaz plan plus Barbados. This oil was intended to satisfy domestic consumption needs only, not to provide for reexports which would compete with Venezuela and Mexico's own exports. Nor was the San José Accord oil intended to subsidize fuel for strictly export industries, as in the case of Jamaica's bauxite, which presumably could afford commercial world market prices. To achieve the 80,000 bpd target during the first quarter of 1981, Pemex would have had to expand deliveries from 13,300 bpd supplied to the beneficiaries when the pact was signed to 80,000 bpd, while Venezuela gradually would diminish exports from the 97,554 bpd registered on August 3 to 80,000 bpd.[14] In fact, volumes for the first three months of 1981 averaged 47,000 bpd for Mexico and 89,000 bpd for Venezuela, as the quest for parity of supply by the exporters took longer than anticipated.[15] The difficulty lay largely in the technical problems encountered in refining Mexico's heavier, more sulfurous crude in simple refineries originally designed to process the Venezuelan cocktail.

The two oil producers promised to grant credits to the importers amounting to 30 percent of the commercial price of their purchases for a period of five years at an annual interest rate of 4 percent. Should the resources derived from these credits finance "economic development projects of priority interest," notably those spurring domestic energy production, the loans could be extended to twenty years at 2 percent interest, with a five-year grace period. These credits were offered at a fraction of the commercial rate; for example, dollar commercial rates exceeded 11 percent at that time, the Eurodollar charge hovered around 10.5 percent, and World Bank loans carried a price tag of 8.25 percent. It was estimated that, at prevailing prices, the nine recipients had a $5,120,000 daily invoice for oil. Thus, the 30 percent credit meant that $1,536,000 would be recycled in the form of low-interest loans.[16]

The donors employed different financing mechanisms. Initially, the purchasing country paid Venezuela the full market price for the crude within sixty days of delivery. Petróleos de Venezuela, S.A. (PDVSA), the national oil company, informed the Venezuelan Ministry of Energy and Mines when the transaction had been completed and the payment received. In turn, the ministry relayed the information to the FIV, which calculated 30 percent of the value of the sale and transferred that amount to the Central Bank, where it was deposited in an account payable to the importer. The latter could either have the money shifted immediately to its own central bank or draw on its account at the Venezuelan Central Bank at a later date. In practice, the FIV made quarterly deposits based on estimated sales during the next quarter. Payments were rendered in dollars, though bolivars would be supplied upon request.

Mexico, whose central bank serves as the official financing institution, followed a much simpler procedure. The beneficiary state merely paid 70 percent of the value of the shipment upon delivery or within sixty to ninety days, retaining the other 30 percent to be repaid as a loan.

Mexico and Venezuela promised to supply equally the needs of recipients, although shipments of petroleum were to be governed by commercial contracts entered into bilaterally by either Mexico or Venezuela and the individual importer, which would have to designate its state oil company or another public entity as its contracting party.[17] Moreover, an effort would be made to dispatch the oil in tankers operated by the Multinational Caribbean Fleet (NAMUCAR), conceived in 1975 to minimize transport costs within the region. The agreement gave rise to government-to-government transactions between Mexico and Venezuela on the one hand, and the importing countries on the other. The foreign firms that operated oil refineries and marketing systems in the beneficiary states expressed no opposition to the joint oil facility, which relieved them of the responsibility for finding, buying, and shipping crude. Generally, these companies received a refining fee, which made their earnings more predictable. Still, the new arrangement diminished their degree of control in the petroleum sector.

Even after the agreement was approved, Venezuela held extensive talks with Trinidad and Tobago in hopes of convincing the island producer to become the facility's third supplier. Eric Williams, Trinidad and Tobago's prime minister, excoriated "Venezuelan imperialism" in the area, and announced the formation of his country's own loan plan for the eleven members of the Caribbean Common Market. Between 1980 and 1983, Trinidad and Tobago pledged $208 million to pay the incremental cost of oil, fertilizer, and asphalt.[18] Economic considerations at home and political problems abroad confined allocations to approximately $75 million.[19]

High-ranking officials from Mexico and Venezuela meet annually, and their subordinates every three to six months, to keep abreast of developments within the facility and to address problems that arise. These coordinating sessions concentrate on technical matters and are attended, from the Mexican side, by representatives of the Central Bank and the Ministries of Foreign Relations, Finance, and Energy, Mines, and Decentralized Industries; and, on Venezuela's part, by representatives of the FIV, and the Ministries of Foreign Relations and Energy and Mines. Other technical experts from PDVSA or one of its operating subsidiaries may participate as the agenda dictates.[20]

Beneficiaries

In light of the economic crisis besetting most of Latin America, why did Venezuela and Mexico limit participation in the San José Accord to the countries of the Central American isthmus, Jamaica, the Dominican Republic, and Barbados—particularly when forswearing ideological and political considerations in selecting participants? To a certain degree, the Economic Cooperation Program for Central American Countries marked a continuation of the Puerto Ordaz agreement, despite conspicuous differences between the two schemes. For that reason, the beneficiaries of the Puerto Ordaz facility were virtually guaranteed membership in the second plan. Besides, these eight nations were traditional Venezuelan customers, seven of which shared both donors' Spanish language, Hispanic heritage, and legacy of impoverishment that attracted the sympathy of decision makers in Caracas and Mexico City.[21] Herrera Campíns obtained the inclusion of Barbados, a small Caribbean state whose modest import needs in 1980 (1,800 bpd) made it a prime candidate for self-sufficiency by mid-decade.

Political and economic considerations militated against membership of other nations in the region. Cuba, which depends on the Soviet Union to provide, on a highly subsidized basis, more than 90 percent of its 200,000-plus bpd consumption, has never sought to affiliate—a prospect diminished by chilled diplomatic relations with Venezuela in the early 1980s. The donors have rejected overtures for membership from Guyana, the Bahamas, Grenada, and from Sheldon Rappaport, a Swiss entrepreneur, who wished to use a refinery in Antigua to supply that and adjacent ministates.

Venezuela lacked enthusiasm for Guyana's admission to the program because the two countries have a messy border dispute involving thousands of square miles of territory believed to be rich in natural resources. In addition, Guyana, which lacks the foreign exchange to pay for oil imports, is not a Caribbean country. The erratically Marxist government of Maurice Bishop, in power from March 1979 until October 1983, made nearby Grenada a political pariah and potential threat in the eyes of the Herrera Campíns administration. Furthermore, while anxious to lend a helping hand to Central America with which it enjoys multiple ties, Mexico has manifested conspicuously less interest in the Caribbean states. It shares Venezuela's belief that the island nations of the basin should seek aid either from Trinidad, their customary oil source, or from Britain, the Netherlands, or France, their mother countries. In addition, the Venezuelan public in general—and opposition political leaders in particular—take a jaundiced view of assisting the English-speaking ministates that have aligned themselves with Georgetown in its frontier controversy with Caracas. Finally, most countries of the region—Guyana and Grenada are examples—do not have refineries and, therefore, could not take advantage of an accord that stipulates the delivery of recon or cocktail exclusively for internal consumption.

In the early phase of the agreement, the donors considered buying or leasing a refinery so that the new mix of crude, which had a growing proportion of heavy and extra components, could be processed for ultimate shipment to participants to satisfy the "growing long term requirements for the countries . . . in the Venezuelan-Mexican Program."[22] In some cases, Mexico and Venezuela have shown flexibility by allowing San José facility crude to be refined in Curaçao. But such instances have been rare because tracking deliveries through several refining centers to beneficiaries has been considered too difficult.[23]

The experience with Haiti confirmed Mexico's and Venezuela's worst misgivings over opportunities for cheating inherent in transshipments. The French-speaking republic gained tentative admission to the facility in late 1980 and received Pemex crude valued at $11 million the following April. According to press reports, U.S. officials believe that two businessmen, one alleged to be then-President Jean-Claude Duvalier's father-in-law, diverted the cargo to Curaçao where it was refined into No. 2 fuel oil and other products.[24] The shift of the seller's market to one favoring buyers foiled the middlemen's plan to reap a sizable profit on the ultimate sale of the shipment, which may have been destined for South Africa. The emerging petroleum glut meant that the fuel oil was worth only $8.4 million at the time, which would still have produced a profit given the 40 percent price break anticipated by Haiti. Outraged at what one observer called "voodoo economics," Mexico billed the Port-au-Prince government for the full $11 million. Both donors subsequently barred the Duvalier government from further aid for failing to live up to its contractual obligations, which included using discounted oil exclusively for domestic consumption. Nevertheless, following Duvalier's ouster in 1986, Mexico and Venezuela stated a willingness to supply Haiti with 6,000 bpd, provided its debt from the previous transaction was paid.[25]

At Mexico's behest, Belize was admitted to the aid program in August 1983 after gaining independence from Britain. The Belmopan government has yet to receive a shipment under the pact, even though Mexico provides products to it on a commercial basis.

Problems

The logistics of transportation and blending posed the first major obstacle to implementing the San José Accord. Most of the region's refineries are small, unsophisticated topping plants—commonly referred to as "coffee pots"—built to accommodate low-sulfur reconstituted Venezuelan crude, not the extra heavy sulfurous Maya grade that Mexico initially insisted should form the bulk of Pemex shipments or even Mexico's lighter but still relatively high-sulfur Isthmus grade.[26] Mexican crude also contains higher levels of nickel and vanadium, metal contaminants that damage costly catalysts in refineries. Even refineries that technically could handle the sulfur and metals in Mexican stock were left with excess heavy fuel oil yields, necessitating the importation of gasoline and diesel oil to meet the pattern of local demand, thereby offsetting—in part—the preferential terms for the crude. The Maya crude also required segregated storage, sometimes necessitating the acquisition of new tanks, which further increased costs. "This is essentially a politically popular deal producing commercially disastrous results," asserted one Caribbean refining expert.[27] Texaco and Esso InterAmerican, which own and operate most of these refineries, were loath to undertake new investment for their adaptation to Maya. Thus, the Dominican Republic, Honduras, and other countries insisted that Venezuela either furnish all of their oil or, at least, pretreat the Mexican crude so that it could be run efficiently through local refineries. For instance, the Dominican Republic's state-owned refinery simply could not handle Mexican crude, prompting ex-President Juan Bosch to condemn the oil deal as "lunacy." To deal with this situation, Venezuela and Mexico embarked upon the arduous task of coordinating the delivery and mixing of their oil. This orchestration, fraught with difficulties, was not accomplished until April 1981, thus explaining the predominance of Venezuela's exports over Mexico's during the first year of the program. In addition, when some countries have produced less gasoline and other so-called white products than needed, Venezuela has reduced below 30 percent the volume of crude in its cocktail shipments to help their refineries attain the desired product mix. Caracas punctiliously notifies the Mexicans whenever such a change is made. Since August 1981, Mexico's willingness to provide chiefly the light Isthmus variety of crude (which is still heavier and more sulfurous than reconstituted oil) to the importers has facilitated the adjustment.

Nevertheless, as late as March 12, 1982, the commission of Mexican and Venezuelan representatives that coordinates the facility received a communication from Honduras declining Mexican oil. This move reflected a protracted dispute between the Honduran government and Texaco, operator of the country's single antiquated refinery that was shut down for a year beginning in September 1981. The U.S. firm resisted Mexican crude on the grounds that its refinery, when operating at full capacity to satisfy domestic demand for light gas and diesel fuel, produced excess fuel oil, which under the terms of the accord could not be exported. Texaco claimed to be losing money because of the agreement, insisting that it would be cheaper to import Saudi Arabian crude directly for use in the country instead of buying the Venezuelan and Mexican crude from the Honduran government. Even though a compromise was hammered out—the refinery operated below full capacity, as Texaco imported 2,000 bpd of product from its Trinidad refinery—Honduras did not resume purchases of Mexican crude until February 1983.

Another problem centered on refining oil delivered under the accord. Financial distress put an end to Mexican-Venezuelan conversations about the possible joint purchase, leasing, or construction of one or more refineries to process their crude for eventual shipment to facility participants and other purchasers.

As I have discussed, expanding the scope of the facility has posed still another challenge to exporting countries. Rather than gradually adding new beneficiaries as originally anticipated, the number of recipients has remained stable, although Belize and Haiti are eligible to receive shipments.

More serious than the membership question has been the recipients' difficulty in designing projects that would qualify for long-term credits. An inability to devise appropriate projects meant that, as of mid-1987, Venezuela had granted only ten of these loans, five to the Dominican Republic and five to Central American countries, compared to seventy under the Puerto Ordaz agreement, while Mexico had not converted a single five-year loan to a long-term credit. Moreover, the Puerto Ordaz program lasted only six years whereas the joint facility was launched in mid-1980.

A U.S. State Department study indicates that three other factors have hampered the development and implementation of projects: absence of guidelines supplied by Mexico and Venezuela; limited institutional capability by the donors to review proposals for long-term financing (a particularly vexing problem for Mexico that lacks an official aid-dispensing agency comparable to Venezuela's FIV); and an acute shortage of cash for expensive, energy-related activities.[28] Inherent in underdevelopment is a shortage of planners, economists, engineers, and other technicians capable of designing proposals that elicit funding. An even more compelling explanation lies in the leaders' preference for proximate instead of long-range solutions in a region where—in the past, at least—coups d'état frequently abbreviate tenure of office. Above all, faced with escalating debt burdens and budget deficits, most governments felt compelled to use their credits for immediate balance of payments relief. The recipient states that have shown the greatest skill in conceiving proposals are Costa Rica, the Dominican Republic, Guatemala, and Panama.

Awarding loans has posed an increasingly difficult challenge to Mexico and Venezuela, both starved for dollars with which to conduct trade and meet payments on their own foreign debt, which exceeds $140 billion between them. Nineteen-eighty-three was a presidential election year in Venezuela, and criticism of foreign aid by business groups and the Confederation of Venezuelan Workers—the nation's major labor organization—at a time of domestic hardship forced temporary suspension of twenty-year Venezuelan credits under the facility in March. Further, PDVSA (anxious to maximize oil income), the FIV (interested in earning higher yields on its loans), and the Central Bank (eager to put the nation's finances in order) had reservations about the program. Consideration of project funding resumed after the annual meeting of the FIV in April 1983; still, the devaluation of the bolivar reduced the resources available for such ventures because most of the fund's loans have been made to the San José countries, and these credits, denominated in dollars, may not exceed 50 percent of FIV's assets. Mexico vowed to continue extending concessionary credits, nothwithstanding the fact that the Central Bank, the Finance Ministry, and Pemex lacked the enthusiasm for the program evinced by the Foreign Ministry, SEMIP, and certain presidential advisers.

Caracas and Mexico City have renewed the San José Accord seven times (1981–1987). The renewals are impressive in the face of the staggering economic problems, including huge foreign debts, afflicting both countries whose foreign exchange earnings suffered because of the worldwide oil surfeit and the attendant fall in energy prices.

In August 1982 the donors agreed to continue granting five-year loans automatically. Meanwhile, they modified the criteria for the twenty-year credits so as to emphasize priority development projects and those that promote regional economic integration. Despite the 50/50 supply provision, Mexico failed to match Venezuela's shipments during the first two years of the facility. Mexican exports averaged only 33,200 bpd, 41.6 percent of its 80,000 bpd target in the first year. Nevertheless, the two suppliers did share the load more equitably during the second and third years. In fact, economic and political considerations in Venezuela have enlarged Mexico's role since August 1982 when PDVSA halted shipments to Nicaragua; yet, during 1985–1986, Venezuela (50,417 bpd) still exceeded Mexico (45,550 bpd) in facility exports,[29] a margin all the more impressive because 11,800 bpd of Mexican exports were on a strictly commercial basis.

Mexico and Venezuela agreed to supply half of each recipient's imports. Nonetheless, refinery conditions and propinquity have dictated Venezuela's status as the exclusive exporter to Barbados. As I have stated, Mexico would be the sole exporter to Belize should this former British colony request oil under the facility. In mid-1982, Venezuela halted shipments to Nicaragua because of the Sandinistas' failure to pay for two consecutive shipments, a situation analyzed in chapter 7. During the fifth year of the pact (1984–1985), Mexico reduced its quotas for Honduras, Jamaica, and Nicaragua, while Venezuela made no deliveries to Nicaragua and increased its quota for Jamaica. Tables 14, 15, and 16 indicate the producers' export levels to each accord member.

Ships for NAMUCAR were never acquired because of the unfavorable economic conditions affecting tanker owners in the 1980s. Hence, the multinational enterprise exists only as a paper fleet.

Economic exigencies led to a hardening of terms in the program's renewal agreement in August 1983. While still guaranteeing up to 160,000 bpd in equal parts, the credit or loan-back portion of recipients' oil bills was reduced from 30 to 20 percent, with the interest for five-year loans doubling to 8 percent and the charge for twenty-year energy and regional integration loans tripling to 6 percent. These changes left the value of concessional aid essentially intact because market prices declined $5 per barrel in 1983. A 20 percent credit on lower prices left the recipient still paying approximately the same dollar amount in cash as with a 30 percent credit on higher prices. In renewing the accord in August 1984, Mexico, reluctant to compound the recipients' economic woes, resisted Venezuela's proposal to raise the interest rates and reduce the credit portion of each transaction to 10 or 15 percent.[30] The two countries did agree to

TABLE 14
Oil Deliveries Under the San José Accord, 1980–1982

	Oil Supplied by Mexico (thousands of bpd)	% of Total	Oil Supplied by Venezuela (thousands of bpd)	% of Total	Total Shipments (thousands of bpd)	Value of Shipments ($ millions/ year)
1980–1981						
Barbados	0	0	1.6	100.0	1.6	$ 4.7
Costa Rica	4.6	44.7	5.7	55.3	10.3	61.7
El Salvador	3.3	29.5	7.9	70.5	11.2	53.3
Guatemala	3.6	27.7	9.4	72.3	13.0	49.8
Haiti	.9	100.0	0	0	.9	11.0
Honduras	.5	8.2	5.6	91.8	6.1	7.2
Jamaica	4.6	23.0	15.4	77.0	20.0	58.8
Nicaragua	5.7	45.2	6.9	54.8	12.6	77.3
Panama	5.2	35.1	9.6	64.9	14.8	66.4
Dominican Rep.	4.8	17.9	22.0	82.1	26.8	43.6
Total	33.2	28.3	84.1	71.7	117.3	433.8
% Volume Contracted		41.6		105.1	73.4	
1981–1982						
Barbados	0	0	1.6	100.0	1.6	6.3
Costa Rica	4.2	40.0	6.3	60.0	10.5	51.4
El Salvador	5.7	46.3	6.6	53.7	12.3	60.6
Guatemala	5.1	44.0	6.5	56.0	11.6	56.7
Haiti	0	0	0	0	0	0
Honduras	0	0	0	0	0	0
Jamaica	7.5	36.1	13.3	63.9	20.8	81.5
Nicaragua	7.1	55.5	5.7	44.5	12.8	86.8
Panama	10.2	45.3	12.3	54.7	22.5	112.1
Dominican Rep.	12.1	46.4	14.0	53.6	26.1	146.6
Total	51.9	43.9	66.3	56.1	118.2	602.0
% Volume Contracted		64.9		82.9	73.9	

Sources: Petróleos Mexicanos: *Petroleum Intelligence Weekly*; Venezuelan embassies in Washington, D.C., and Mexico City.

restrict the volume pledged to 130,000 bpd, an amount that still exceeded actual deliveries in previous years. At the instance of Venezuela—a nation anxious to advance the Contadora process in Central America—the donors committed themselves to terminating oil-related aid to any country initiating "warlike" activities. This proviso appealed to both parties: the Mexicans, well known for their pro-Sandinista sympathies, could argue that it would prevent a Honduran invasion of Nicaragua, while the Venezuelans might contend that it would deter incursions by the Managua regime into its neighbors' territory.

The most important change in 1984 found Venezuela stipulating that half of

TABLE 15
Oil Deliveries Under the San José Accord, 1982–1984

	Oil Supplied by Mexico (thousands of bpd)	% of Total	Oil Supplied by Venezuela (thousands of bpd)	% of Total	Total Shipments (thousands of bpd)	Value of Shipments ($ millions/ year)
1982–1983						
Barbados	0	0	1.3	100.0	1.3	$ 2.8
Costa Rica	4.3	44.8	5.3	55.2	9.6	48.6
El Salvador	5.6	52.3	5.1	47.7	10.7	62.4
Guatemala	5.6	46.7	6.4	53.3	12.0	59.8
Haiti	0	0	0	0	0	0
Honduras	.9	15.5	4.9	84.5	5.8	9.8
Jamaica	5.8	25.8	16.7	74.2	22.5	48.6
Nicaragua	11.3	100.0	0	0	11.3	129.9
Panama	11.5	50.4	11.3	49.6	22.8	123.9
Dominican Rep.	3.1	50.4	12.9	49.6	26.0	137.8
Total	58.1	47.6	63.9	52.4	122.0	623.0
% Volume Contracted		72.6		79.9	76.3	
1983–1984						
Barbados	0	0	.7	100.0	.7	1.1
Costa Rica	5.4	43.2	7.1	56.8	12.5	57.7
El Salvador	6.4	43.5	8.3	56.5	14.7	67.8
Guatemala	4.3	100.0	0	0	4.3	44.0
Haiti	0	0	0	0	0	0
Honduras	1.9	23.5	6.2	76.5	8.1	20.0
Jamaica	1.9	14.8	10.9	85.2	12.8	19.3
Nicaragua	5.8	100.0	0	0	5.8	61.6
Panama	11.0	43.8	14.1	56.2	25.1	109.0
Dominican Rep.	13.9	46.3	16.1	53.7	30.0	141.6
Total	50.6	44.4	63.4	55.6	114.0	522.1
% Volume Contracted		63.3		79.3	71.3	

Sources: Petróleos Mexicanos: *Petroleum Intelligence Weekly*; Venezuelan embassies in Washington, D.C., and Mexico City.

TABLE 16
Oil Deliveries Under the San José Accord, 1984–1987

	Oil Supplied by Mexico (thousands of bpd)	% of Total	Oil Supplied by Venezuela (thousands of bpd)	% of Total	Total Shipments (thousands of bpd)	Value of Shipments ($ millions/ year)
1984–1985						
Barbados	0	0	.6	100.0	.6	$ 1.4
Costa Rica	0	0	9.5	100.0	9.5	58.6
El Salvador	5.7	44.9	7.0	55.1	12.7	69.4
Guatemala	6.8	46.0	8.0	54.0	14.8	0
Haiti	0	0	0	0	0	25.7
Honduras	2.5	29.4	6.0	70.6	8.5	31.1
Jamaica	(3.1)	22.5	10.7	77.5	13.8	26.6
Nicaragua	2.6	100.0	0	0	2.6	93.8
Panama	9.3	62.0	5.7	38.0	15.0	116.3
Dominican Rep.	13.8	51.0	13.5	49.0	27.3	0
Total	43.8	41.8		58.2	104.8	
% Volume Contracted		67.4		93.9	80.6	
1985–1986						
Barbados	0	0	.7	100.0	.7	$ 1.7
Costa Rica	(3.1)	31.6	6.7	68.4	9.8	18.3
El Salvador	7.1	51.8	6.6	48.2	13.7	54.2
Guatemala	(5.8)	49.2	6.0	50.0	11.8	47.6
Haiti	0	0	0	0	0	0
Honduras	1.4	28.6	3.5	71.4	4.9	7.3
Jamaica	(2.9)	27.6	7.6	72.4	10.5	25.3
Nicaragua	0	0	0	0	0	0
Panama	8.5	57.4	6.3	42.6	14.8	74.1
Dominican Rep.	16.7	55.9	13.2	44.1	29.9	104.2
Total	45.5	47.3	50.6	52.7	96.1	332.7
% Volume Contracted		70.0		77.9	73.9	
1986–1987						
Barbados	0		n.a.		0	0
Costa Rica	1.7		n.a.		1.7	5.6
El Salvador	5.1		n.a.		5.1	13.7
Guatemala	4.1		n.a.		4.1	10.7
Haiti	0		n.a.		0	0
Honduras	0		n.a.		0	0
Jamaica	1.4		n.a.		1.4	1.4
Nicaragua	(.6)		n.a.		.6	2.9
Panama	8.6		n.a.		8.6	21.1
Dominican Rep.	18.1		n.a.		18.1	54.0
Total	39.6		n.a.		39.6	109.4
% Volume Contracted		60.9		n.a.	60.9	

Sources: Petróleos Mexicanos: *Petroleum Intelligence Weekly*; Venezuelan embassies in Washington, D.C., and Mexico City.

1. Bracketed figures denote exclusively commercial sales.

2. August 1986 to March 1987.

its San José credits would be in nonconvertible bolivars. In light of the unfavorable dollar-bolivar exchange rate following the bolivar's 1983 devaluation, this move encouraged the purchase of Venezuelan goods and services by beneficiary states. In August 1986, Venezuela denominated its credits solely in dollars, but with the proviso that the monies could be used only for buying goods and services of Venezuelan origin.

Mexico's credits have always been earmarked for purchasing that country's products. In 1985 Mexico initiated a further tightening of terms, insisting on 100 percent payment for crude supplied; however, recipients were permitted to draw funds equivalent to 20 percent of the oil price for "qualifying energy development projects" from a special $72 million credit within the Central Bank of Mexico.

In renewing the pact in August 1986, Presidents de la Madrid and Jaime Lusinchi, Campíns's successor, commented on the "instability" and "uncertainty" of the international petroleum market whose condition had so adversely affected the economies of their respective countries. Nevertheless they stressed the "political will . . . to support cooperative actions that contribute to the economic and social development of the region such as the guarantee of petroleum supplies and the allocation of critical resources to finance regional trade"—an unmistakable reference to the effort to promote Mexican and Venezuelan nonpetroleum exports through the San José Accord. At the same time, the chief executives reduced from twenty to twelve years the period for long-term loans under the program.[31] This move was largely an academic exercise because of the shortage of funds for long-range purposes. Part of the reason may have been that the recipients did not want to use the credits for long-term loans, preferring instead to channel the funds to short-term balance-of-payments support. Mexico has not granted developmental loans, and in May 1986 Venezuela cited economic hardship in halting disbursements for these ventures. Beneficiaries owed $521 million in long-term loans, and Caracas could ill afford to dispense even more largesse.[32] No major changes in the accord were made at the 1987 renewal.

Both Mexico and Venezuela have exaggerated the amount of aid bestowed under the joint facility. Former President López Portillo was especially prone to hyperbole. In his sixth State of the Nation address, he claimed that Mexico alone had made $700 million available to beneficiaries between 1980 and 1982. In fact, as of August 1986, accumulated value of oil delivered under the accord exceeded $3 billion as financing totaled $610.3 million, distributed as follows: Dominican Republic, 29.0 percent; Panama, 23.7 percent; El Salvador, 14.4 percent; Guatemala, 11.4 percent; Costa Rica, 10.2 percent; Jamaica, 8.9 percent; and Honduras, 2.4 percent. At the same time, overall indebtedness amounted to $267.4 million or 43.8 percent of total financing, of which slightly more than one-quarter had been renegotiated with different terms and for varying periods.

Neither donor has undertaken a cost-benefit study to determine the precise value of the facility. On one hand, such an analysis would concentrate on the monies saved by importers thanks to the loan-back scheme. On the other hand, it would consider—inter alia—expenses entailed in: (1) modifying refining capabilities, (2) segregating Mexican crude and Venezuelan reconstituted oil in storage facilities, (3) coping with delays entailed both in coordinating deliveries and blending the two varieties, (4) importing supplementary products when refinery runs of heavier stock produce more fuel oil, and less gasoline and diesel, than is consumed locally, (5) foregoing opportunities—once locked into contractual agreements with the donors—to buy on the spot market, and (6) purchasing goods and services from Venezuela and Mexico rather than on the open market.

Arguably, less-developed countries would profit more from an aid scheme such as the San José Accord if they enjoyed access to products as well as crude imports, could sell abroad surplus products derived from imported crude, and were allowed to swap heavy for light crude (and vice versa) with third parties, depending upon internal refining conditions. Still, the donors have exhibited remarkable flexibility: (1) by their willingness to exchange light and heavy oil to satisfy local refinery configurations and by occasionally supplying products; (2) by one partner's adjusting shipments to compensate for changes in deliveries by the other exporter; and (3), when necessary, by one supplier's meeting the total needs of a given recipient.

Initially, the San José Accord's terms combined with the promise of a secure supply of oil, made the agreement an appealing aid mechanism for the region's developing countries. For that reason, along with a desire to involve the two donors more actively in regional affairs, the Reagan administration was eager to pull the program under the umbrella of the Caribbean Basin Initiative and to insist, disingenuously, that Mexico and Venezuela were "sponsoring countries" of this highly publicized U.S. assistance venture in the area.[33]

Involvement in the San José Accord has proved a learning experience for both donors, particularly Mexico, which entered the program during the heyday of an ideological foreign policy designed to advance its status as a Regional Leader. For instance, both countries discovered that—the proclaimed desirability of divorcing aid and politics aside—it was impossible to ignore political factors in granting assistance. A good-faith effort was made by Venezuela's Christian Democratic administration (1979–1984), which aided the Sandinistas whom it often criticized. For its part, Mexico supplied oil to the governments of Guatemala and El Salvador, despite refugee and border problems with the former, as well as protests mounted by left-wing opposition groups in Mexico City accusing the government of subsidizing repressive regimes in these countries without imposing any conditions.[34] Yet, as I will discuss in chapter 7, both Venezuela and Mexico felt a common need to draw the line and terminate or curtail oil shipments to Nicaragua because of the Managua regime's failure to pay its bills. As noted, different political considerations lay behind the August 1984 sanction on "warlike actions" by recipients. Moreover, Guyana's absence from the facility is as much due to its Essequibo border dispute with Venezuela as to its lack of appropriate refining facilities. Until the October 1983 overthrow of the radical New Jewel Movement, Caracas was also wary of Grenada's regime. The absence of a refinery on the postage stamp–sized island afforded a justification for excluding it from a crude-export facility, although this is not always a deciding factor in eligibility criteria (for example, Belize also lacks a refinery).

Even more frustrating is the paucity of political influence derived from the program. Actually, one ranking Venezuelan diplomat, who asked to remain anonymous, confided that "dealing with the demands and complaints of ten aid recipients had made us more sympathetic to the problems that the United States faces in its relations with the Third World. The 'what-have-you-done-for-me-lately' approach of certain countries has led us to sympathize with the plight of foreign aid donors," he added. Privately, some Mexican leaders have voiced the same sentiment. These officials are individuals identified with the liberal-rationalist perspective, for nationalist-populists in the Foreign Ministry and elsewhere have continued to champion Mexican participation in the program. The members of the accord, in particular Panama and the Central American states, evince smoother relations with Venezuela than Mexico, probably as a result of long-term oil trade with the former and refineries tailored to Venezuelan oil. Central American wariness toward Mexico may also reflect political discomfort on the part of their regimes with both Mexico's intrusiveness in the area and its revolutionary nationalist dogma.

As it adjusts to the role of a Responsible Debtor, Mexico has shown far less forbearance with its own debtors, some of which now shun Pemex oil to buy in the spot market. After all, Mexico must attempt to meet its international obligations and expects the same from the beneficiary states. Furthermore, devoting even 2 percent of its total production to foreign assistance poses problems during a period of austerity. Belt-tightening in both Mexico and Venezuela has led to the modification of the facility, to the point where new long-term loans have disappeared. Aggravating the problem of recipients is an inability to devise proposals suitable for long-term assistance. The upshot is that, while furnishing some badly needed balance-of-payments support, the San José Accord has not stimulated energy development by the beneficiaries.

In view of the diminishing economic justification for the San José Accord in a world awash in oil, might the program be terminated? For several reasons the answer is no, although the volumes will decline until prices rise substantially. To begin with, the exporters need all the contract customers they can get in view of competition with the attractive spot market. In addition, resources do flow to poorer countries that run chronic balance-of-trade deficits with the donors, even though Mexican and Venezuelan exports generated by the pact have been modest. Admittedly, the scheme does contribute marginally to the economic well-being of impoverished countries in a strife-afflicted area. Above all, Mexico and Venezuela, like it or not, have undertaken a commitment from which it will be difficult to disengage themselves; its termination would spark angry denunciations of both nations which, current problems aside, believe they derive an element of prestige from the pact. Still, leaders in both countries have minimized their idealistic rhetoric about the accord as they pragmatically pursue their own self-interest in dispensing aid. They appear more aware of the limited impact that money has in overcoming structural and cultural impediments to change.

7

Mexico and Nicaragua

The oil-impelled international activism of López Portillo manifested itself most clearly in Mexico's dealings with Nicaragua. Mexico's alleged commitment to nonintervention in the affairs of other states aside, the López Portillo administration reviled General Anastasio Somoza Debayle, overtly aided the Sandinista rebels, and deftly foiled U.S. efforts to accomplish a gradual transition to a moderate regime following the dictatorship's collapse. Furthermore, Mexico's support for the insurgents continued after their seizure of power in July 1979. López Portillo staked his claim to regional leadership by furnishing generous assistance to the new Government of National Revolution, most notably with oil supplied by Pemex. Yet growing pragmatism in Mexico City, engendered by the 1981–1982 oil glut cum financial crisis, produced a major shift in bilateral energy relations—as the Soviet Union supplanted Mexico as the principal source of oil for Managua's beleaguered revolutionaries.

Mexican-Nicaraguan Relations

Throughout its history, Mexico has enjoyed closer ties to Nicaragua than to any other Central American country. This affinity springs from strong mutual interests in preventing a unified Central America under the tutelage of Guatemala, a neighbor whose relations with Mexico have been strained by a mistrust born of inequality in size, population, and wealth. When Guatemalans decry the "colossus of the north," they are referring to Mexico, not to the United States. The most recent sources of controversy have been: Guatemala's territorial claims to Belize, which Mexico adamantly rejects; the influx of more than 100,000 Guatemalan refugees into Chiapas, Mexico's southernmost state; and the violation of Mexican sovereignty by Guatemalan military units chasing guerrillas believed cached in Chiapan refugee camps.

Despite their common distrust of Guatemala, the paths of Mexico and Nicaragua diverged sharply in the 1930s. In Mexico, Cárdenas authored reforms to improve the social and economic plight of workers and peasants, while broadening participation in the national revolutionary party, later to become the PRI. In Nicaragua, Anastasio "Tacho" Somoza García, the cunning, congenial fair-haired boy of the United States Marines who had for nineteen years (1912–1925 and 1927–1933) occupied the Alabama-sized nation, began to erect a political structure that channeled resources to a small elite composed of his family and their loyalists. Thanks to the devotion of the Marine-trained National Guard which he headed, Somoza could boast that the "three Ps" encapsulated his governing philosophy: "Al amigo, le doy plata; al indiferente, le doy palo; y al enemigo le doy plomo" ("To friends, I give money; to neutrals, I give the stick; and to enemies, I give lead or bullets"). An early victim of this philosophy was Augusto César Sandino, a Liberal politician and avid nationalist who had led guerrilla raids against the U.S. forces during their second occupation. In an act of treachery, Somoza ordered the National Guard to capture and execute his potential rival in 1934.

Somoza got a dose of his own medicine on September 21, 1956, when Rigoberto López Pérez, a young Nicaraguan poet, used *plomo* to assassinate the venal, swag-bellied dictator. Nonetheless, Tacho had made provisions for a smooth political transition and, upon his death, his son Luis donned the presidential sash.

Like his father, Luis did not hesitate to use the National Guard to consolidate his power. Still, he increased public spending on health, welfare, education, housing, and agrarian reform. He also reduced political repression, modernized the Liberal party dominated by his family, and secured a constitutional amendment to prevent another Somoza from succeeding him as chief executive. The changes were short-lived, for in 1967 Luis's brother, Anastasio "Tachito" Somoza Debayle, assumed the presidency after a blatantly corrupt election. In the view of political scientist Thomas W. Walker, the country passed from a period of tentative optimism into a venal dictatorship.[1]

A series of events in the 1970s catalyzed opposition to Somoza and eroded the legitimacy of his regime. He and the National Guard turned the December 1972 earthquake that devastated Managua to their pecuniary advantage. They commandeered international relief aid, which they then sold to victims of a tragedy that took 10,000 to 20,000 lives. They realized huge profits speculating in land on which the homeless were eventually resettled. And they awarded contracts to Somoza-controlled companies to build a new road network, the cost of which escalated because of the exorbitant prices charged by the firms. Such corrupt, self-enriching acts nourished the belief that Somoza had gone too far. Even businessmen chafed at the injustice of shouldering new emergency taxes while the strutting, preening Tachito avoided paying taxes even as he brazenly squandered public funds.

Recourse to the "three P's" further undercut Somoza's standing. He was linked in the public's mind to the brutal murder of Pedro Joaquín Chamorro, an internationally renowned journalist who edited and published *La Prensa,* Nicaragua's leading newspaper, and who was an inveterate critic of the regime's peculations. Under martial law, the National Guard overreacted to highly publicized initiatives of the Sandinista guerrillas—with beatings, torture, rape, arbitrary imprisonments, and summary executions. The opposition grew in proportion to the savagery of Somoza's praetorian forces—until civil war pitting the National Guard against the FSLN engulfed the country.

As I have discussed in chapter 2, oil wealth encouraged López Portillo to act as a Regional Leader with respect to Nicaragua. He worked sedulously to undermine Somoza, and—following the dictator's overthrow—extended diplomatic and economic support to the revolutionary regime. The most ambitious aid came in the form of oil. Not only did Mexico agree to supply Nicaragua with 7,500 barrels per day, but also it promised to assist Nicaragua in "all aspects" of the hydrocarbon industry, especially in exploration and drilling. Further, Mexico donated two geothermal well-drilling rigs and pledged to train the personnel needed to operate them. In return for this help, the Nicaraguans agreed to sell Mexico hides, meat, sugar, oil seeds, gold, and precious stones.[2]

In April Venezuela, which had sold oil on highly favorable terms to the Somoza regime under the 1974 Puerto Ordaz agreement, also committed itself to supplying $55 million in oil to Nicaragua at 4 percent interest, with a five-year grace period and the possibility of converting the aid to long-term loans for industrial development at 2 percent interest. Venezuela's Energy Minister Humberto Calderón Bertí expressed satisfaction that Mexico had begun sharing responsibility for providing crude to Nicaragua. Mexico and Venezuela stitched their aid to Nicaragua into the San José Accord, unveiled in August 1980.

Nicaragua's Energy Picture

Mexico's foreign energy assistance to the Sandinistas was especially important because Nicaragua produced no oil of its own. The late 1960s and early 1970s saw intense exploration activity in Nicaragua, the largest of the five Central American nations. Esso, Elf, Erap, Chevron, and Union initiated offshore seismic surveys, while Shell, Superior, and Union were among the firms that drilled exploration wells—with the greatest attention lavished on areas off the Pacific and Caribbean coasts. The 1965–1975 period witnessed the completion of three wildcat wells onshore south of Managua, two wells along the northern part of the Caribbean coast, six wells offshore in the Pacific, and twenty-two wells offshore in the Caribbean.[3]

None of these ventures proved fruitful, even though Esso found small quantities of gas in its Pacific concession in the late 1960s, and El Paso discovered noncommercial volumes of oil in its Huani-1 Caribbean well in 1974. Subsequent efforts included test wells drilled by Chevron (1975), Occidental (1976), Union (1978), and Texaco/Amerada Hess (1978).[4] The rise to power of the Sandinistas in mid-1979 brought economic uncertainties to the country, a political environment increasingly hostile to private enterprise, and—after months of unsettling speculation about its content—promulgation of a new investment law (1982) whose profit-sharing provisions were ambiguous and unattractive. Although the Oceanic Exploration Company reportedly began negotiating for renewal of its 4.36-million-acre offshore contracts,[5] the general response was a halt in field activity: Texaco/Amerada Hess, Western Caribbean Petroleum/Occidental, and Frank Petroleum/Pyramid Oil relinquished their concessions; groups headed by Buttes, Phillips, CMMSA, Oceanic, Rosario Mining, and Union suspended activities; and Petróleos de Nicaragua (Petronic), the state oil firm created in 1983, lacked resources with which to finance exploration.[6]

Disappointing results in the hydrocarbon field sparked interest in developing geothermal resources and acquiring alcohol technology. Three active geothermal wells at Momotombo, southwest of Lake Managua, were producing eighteen megawatts at the end of 1981, with Foramines scheduled to drill four more wells the following year.[7] Later, the Italian and Canadian governments lent assistance to the geothermal and hydroelectric project, coordinated by the Inter-American Development Bank, but the results have been modest. All told, geothermal sources satisfy approximately 20 percent of national electricity needs. In May 1985 Italy committed itself to help Nicaragua build a second unit at the Patricio Arguello Ryan geothermal power plant near the town of La Paz Central. The existing unit, to which Italy made substantial contributions, saves Nicaragua $50,000 per day in energy costs, according to the Italian ambassador in Managua.[8] In June 1981 a group of Nicaraguan officials visited Brazil to negotiate an exchange of oil exploration rights to be exercised by Braspetro, a Brazilian state firm, in exchange for Brazilian alcohol know-how. Nothing had materialized from this contract as of late 1987.

In the early 1970s, the Compañía Marítima Nacional, headed by President Somoza, sought to raise $300 million for a proposed Monkey Point refinery, with its initial capacity of 250,000 bpd to grow to 1 million bpd supplied with Mideast crude through a projected offshore terminal capable of accommodating 500,000-ton tankers. It was also suggested that Monkey Point could serve as a transshipping facility for the unloading at a Pacific port of Alaskan crude, which would then be pumped to the Atlantic because large crude carriers cannot transit the Panama Canal.[9] The project was abandoned because of a dearth of investment capital and a surfeit of political turmoil.

Consequently, Refinería Esso Managua, an Exxon subsidiary, operates the country's only refinery. The facility is located in Managua and joined by a thirty-nine-mile-long pipeline to the Pacific port of Puerto Sandino (formerly Puerto Somoza). Constructed in 1963 with a 6,000 bpd maximum capacity, the topping plant's "throughput" was enlarged in 1971 to 18,500 bpd to satisfy growing internal demand. Its rated capacity in 1987 was 15,000 bpd. Esso Standard Oil, S.A., Ltd. handles overall operations in the country.

Mounting Problems

As mentioned above, the Sandinistas found themselves swimming in oil within months of seizing power—thanks to the generosity of Mexico and Venezuela. Despite initial problems related to both the quality of petroleum and the coordination of deliveries, the San José Accord worked satisfactorily until mid-1982. At that time, Venezuela suspended shipments to Nicaragua because the GNR missed payments on two shipments, totaling 500,000 barrels and valued at $18.7 million. Persistent requests by the Sandinistas to allow them to stretch out repayments over an extended period did not move Venezuelan politicians, who faced ever more intractable economic problems as the result of the worldwide oil glut. As noted in chapter 6, 1983 was an election year in Venezuela, and labor and business groups complained about dispensing aid abroad at a time of serious financial difficulties at home. In addition, PDVSA withdrew from joint oil exploration with its Brazilian counterpart in Nicaragua's Pacific coast waters, thereby forcing indefinite postponement of the project. Shigeaki Ueki, president of Petrobras, Brazil's national oil company, said that prospects for discovering oil were encouraging in Nicaragua but his country lacked the investment monies necessary to pursue the venture.

On the heels of Venezuela's suspension of shipments, Mexico pledged to meet Nicaragua's needs. While not agreeing with every Sandinista policy, Mexico City wished to project influence in the region, evince independence of Washington, and moderate the actions of—if not coopt—the leftists in Managua.

Mexico's attentiveness to the Sandinistas was obvious when its spokesmen criticized sharply the bombing of Nicaraguan oil facilities by the anti-Sandinista contras, who enjoyed the CIA's covert support. In a September 1983 incident, the insurgents bombed the undersea oil pipeline, hose, and buoys leading to the Esso refinery, and, the next month, demolished a storage tank containing diesel fuel in Corinto, the port 68 miles northwest of Managua that receives imports of petroleum products. These commando-style attacks, which failed to destroy stocks of aviation fuel crucial to crop dusting the cotton fields that produce the country's second most important export, led to a change in deliveries of Mexican crude as Exxon refused to dispatch its tankers into an area declared a "war zone" by the contras. Pemex was wary about using its own vessels as rumors circulated that at least one Mexican crew had demanded to return to port upon learning that its ship was bound for Nicaragua. Nonetheless, shipments continued after Exxon helped Petronic charter tankers from other owners. The attack came at the most crucial time of the year because of the impending coffee harvest that generates 80 percent of the nation's foreign exchange.[10] According to a Pemex spokesman, Mexico and Venezuela provided technical assistance to help repair storage tanks sabotaged in the Gulf of Corinth.[11]

Mines laid around Nicaragua harbors by the contras damaged at least twelve ships, including a Soviet oil tanker. On March 20, 1984, the *Lugansk* struck a mine while carrying 250,000 barrels of Soviet crude into the Pacific coast port of Puerto Sandino. The CIA-supplied device ripped a hole in the hull and injured five crewmen. Eight days later, the Liberian-registered *Iver Chaser* was damaged while leaving Corinto with a mixed cargo of molasses and benzine. The presence of the mines disrupted shipping to and from Nicaragua, even though most vessels managed to load or discharge their freight. At the time of the *Lugansk* incident, two other Soviet tankers containing refined petroleum were awaiting a secure entry into the port.[12]

Mexico excoriated the insurgency. Even more strident than his colleagues in Mexico City, Miguel Marín-Bosch, a Mexican U.N. diplomat, urged the Security Council to hear Nicaragua's complaint against the mining of its ports and interruption of free navigation in its waters, while reminding the body of the U.N. resolution, adopted in October 1983, which condemned interference with navigation by belligerents in the Iran-Iraq War.[13] On April 13, 1984, Foreign Minister Sepúlveda denounced the mining of ports and called for the total elimination of all armed violence, direct or indirect, against Nicaragua.[14] Muñoz Ledo, Mexico's extremely nationalistic U.N. ambassador, embellished this theme when he censured the U.S. trade embargo on Nicaragua as violating "one of the fundamental principles of international law"—that is, the tenet "prohibiting [the use of] coercive economic measures."[15]

The Minuet

Mounting economic challenges at home, including spiraling prices, massive unemployment, anemic private investment, and a mounting foreign debt, prompted Mexico to reconsider its virtual carte blanche for oil purchases extended to Nicaragua. This reassessment came despite the regime's promulgating an austerity program in 1985 that involved stimulating the private agricultural sector, reining in government spending, restraining real wages, and raising prices of gasoline (150 percent), diesel fuel (225 percent), bottled gas

(220 percent), and electricity (50 percent). At this point, officials in the Office of the Mexican Presidency, the Ministries of Foreign Relations and Energy, and the PRI generally favored continued assistance to Nicaragua for political reasons already mentioned. Left-of-center parties and newspapers championed even more aid to Managua. In contrast, those entities where liberal-rationalists abound—the Finance Ministry, Central Bank of Mexico, and Pemex— viewed with a jaundiced eye the provision of oil gratis or in return for nominal payments. Moreover, contrary to expectations, Mexican diplomats had made little or no headway in curbing the Sandinistas' heavy-handedness toward such critics as the Roman Catholic hierarchy, *La Prensa*, the business sector, and opposition parties. "It's like a husband's paying the hotel bill for the trysts of an unfaithful wife," a Mexican congressman observed.

The growing influence of the liberal-rationalists gave impetus to a kind of minuet between the two countries. A Mexican official took the first step by publicly chiding the Nicaraguans for not meeting their contractual obligations. A suspension in oil shipments then followed. However, Mexican officials were careful to reiterate their support for the Sandinista revolution and assure their friends in Managua that exports would resume as soon as they paid off the arrears. Next came a bow from President Daniel Ortega Saavedra or another high-ranking *comandante*, who jetted to Mexico City to promise that his country would live up to its word. A gesture—perhaps a token payment— typically accompanied this move. With a flourish, the Nicaraguans made an overture to Mexican nationalism by declaring that assistance was needed to accomplish revolutionary reforms in the face of "Yankee aggression and imperialism." Mexico completed the minuet by taking a half-step backwards; namely, after halting exports, it declaimed both its enduring fraternal relationship with the Sandinistas and its readiness to continue shipments "for as long as necessary"—however, at a lower level and on more rigorous terms.

Since 1983 there have been repeated performances of the Mexican-Nicaraguan minuet, although slight variations have enlivened each rendition. On February 1, 1983, Vice Finance Minister William Hupper led a Nicaraguan delegation to Mexico that met with a group of Mexican officials headed by Navarrete, Mexican undersecretary of foreign relations for economic affairs. Although the Mexicans did not publicize the possibility that Pemex might curtail deliveries to Nicaragua, Hupper publicly remarked that any aid to finance Nicaraguan energy bills was "very important" because Nicaragua was suffering from a drop in the prices of its own exports.[16] In June and July 1983, Mexico held up supplies to pressure Nicaragua to renegotiate its $300 million debt. Washington admitted to being more than a casual observer of the political dance. "We have pressured Mexico because we feel that Nicaragua is not the type of government which deserves this kind of financing," a U.S. diplomat said. "It would not break our hearts to see them go bankrupt," he added.[17] Nevertheless, on August 23 the minuet was completed after Nicaraguan Foreign Minister Miguel d'Escoto Brockman and Planning Minister Henry Ruiz flew to Mexico City aboard an aircraft provided by de la Madrid, for a trip described by the visitors as "routine, normal, and important." Following a meeting with de la Madrid and Sepúlveda, Ruiz claimed that petroleum supplies were "assured," for Mexico had granted terms and facilities to enable Nicaragua to pay its debt in accordance with its means.[18] Ruiz's statement excited editorial praise from *El Nuevo Diario,* a pro-Sandinista daily in Managua: "The news is so good that it implies yet another defeat for Nicaragua's enemies. Their attempts to get the Mexican Government to end its daily supply of 15,000 barrels of oil [sic] which we use to run our factories and our country's vehicles, have been defeated."[19]

Euphoria in Managua turned to chagrin in January 1984 when Mexico again suspended deliveries because of Nicaragua's failure to satisfy the terms of the San José Accord. Nicaragua responded on February 11 by sending Petronic Director José D. Gómez to visit Pemex's massive refining and petrochemical installations in Coatzacoalcos. After his tour, Gómez reported that Mexico would continue to dispatch 7,500 bpd of crude oil to Nicaragua from the Pacific port of Salina Cruz. In addition to oil, he expressed interest in obtaining liquefied gas from Pemex. The Mexican government completed the minuet on February 22, 1984, by announcing officially that petroleum shipments would be renewed because Nicaragua had agreed to pay $8 million in overdue debt.[20]

In January 1985 Commerce Secretary Héctor Hernández Cervantes lamented Nicaragua's failure to pay for oil imports, even when Mexico agreed to accept meat, cotton, or other bartered goods in lieu of dollars for half of the 80 percent cash payments under the San José Accord. "If they can't give us the goods," he said, "they'll have to give us dollars. It's up to the Nicaraguans to expand their trade with us. Our crude exports won't pass a certain level [this] year."[21] Hernández did not cite the level he had in mind.

In late February 1985, Deputy Energy Secretary Eliseo Mendoza Berrueto announced that Mexico had "temporarily suspended" shipments to both Costa Rica—pending payment of a $15 million debt (the San José government also owed Mexico for technical assistance, drilling, and exploration done by Pemex)—and to Nicaragua, whose debt to Mexico exceeded $600 million for oil alone.[22] Nicaragua's ambassador to Mexico quickly called a press conference, saying that the oil cutoff "has no political implications."[23] This conclusion was naive because the United States in fact had been encouraging de la Madrid's government to assume a tougher stance vis-à-vis the Sandinistas. "It is not usually done in a direct manner," said a State Department official. "We deal more effectively with Mexico indirectly."[24] Although a tanker carrying

250,000 barrels of Pemex oil reached Puerto Sandino on March 21, 1985, the Mexican government stressed that it was responding to an "emergency situation" and that future deliveries would depend upon Nicaragua's clearing up its unpaid debts.

On April 8, both countries entered an accord broaching the prospect of barter deals, which might absorb up to 50 percent of Managua's debt. "There is no question of cutting the shipments," claimed a diplomatic source at the Managua meeting. One solution contemplated was giving Mexican fishing trawlers access to Nicaragua's Atlantic waters or permitting exploitation of valuable timber holdings on the same coast.[25]

Apparently, the Nicaraguans expected continued Mexican forbearance. In any case, high-level officials did not react to the announced cutoff until Ruiz visited Mexico in mid-April. Reportedly, de la Madrid himself broke the news to Ruiz that there would be no further deliveries unless Managua paid 80 percent in cash in advance. The message, according to Deputy Foreign Minister Víctor Hugo Tinoco was: "We're not in a position to supply any more oil unless you pay cash for the oil that you receive."[26]

The imposition of a U.S. economic embargo on Nicaragua in early May 1985 precipitated a political firestorm in Mexico. Epitomizing the outcry was an article in the left-leaning newspaper *Universal,* entitled "Petroleum to Nicaragua; PEMEX Lies." Despite official claims that Pemex was shipping oil to Sandinistas, the author wrote, "Nicaragua is not even receiving a toothpick from Mexico." Furthermore:

> This is the whole truth about the government's behavior toward the harassed Sandinista revolution. However, it is clear that the Mexican Government is trying to deceive the public to hide the incongruence of the effort that has resulted from supporting the initiatives of the heterogeneous Contadora Group with the refusal to sell oil to that Central American country when it needs it the most.[27]

In the face of this and similar diatribes, Mexico agreed to make more oil available under the "norms of the San José Pact." This pledge was given by Eduardo Pesqueira Olea, Mexico's secretary of agriculture and water resources and a de la Madrid confidant, who headed Mexico's delegation to the first meeting of a Mexican-Nicaraguan Mixed Commission, convened in Managua May 27 to May 29, 1985. In addition to a June shipment of 250,000 barrels, he expressed Mexico's "disposition" to furnish 320,000 barrels of crude between July and September and another 410,000 barrels during the last quarter of 1985. Nicaragua would have to pay for this oil with foreign exchange, he added. The commission also discussed technical cooperation in agriculture, forestry, and livestock, as well as Mexican export opportunities in chemicals, replacement parts for agricultural machinery and vehicles, and other items in the wake of Washington's curtailing trade with Nicaragua. Mexican assistance loomed all the more important because in mid-1985 Honduras, which supplied surplus electricity to Nicaragua from its El Cajón generating facility, threatened to halt deliveries unless the government paid its May bill totaling $176,000. Nevertheless, during the remainder of 1985, Mexico Pemex did not ship the 730,000 barrels cited by Pesqueira because of Managua's chronic shortage of foreign exchange and its decision to rely more heavily on Russian oil. Although not publicized in Mexico, Pemex reported making shipments that totaled 100,000 barrels of Maya crude in late 1986 and early 1987. Reportedly, these exports, which would help meet Nicaragua's asphalt needs, were on a cash basis.

Russians to the Rescue

Diminishing Mexican supplies compelled the Sandinistas to cast about for alternative sources of crude and products. In April 1981 Libya lent Nicaragua $100 million for agricultural development and reconstruction of earthquake and war damage. Rumors notwithstanding, Colonel Muamar Quadaffi has yet to dispatch the first tanker to Puerto Sandino. Like Libya, Iran made inflated promises that have not panned out. Vagueness characterized a May 1982 bilateral deal between Nicaragua and Iran. Presumably, the former would export sugar, meat, cigarettes, and coffee in return for the latter's petroleum. According to Manuel Cordero Cuadra, a minister-counselor in Nicaragua's Washington embassy, Iran had made available at least one tanker, the contents of which were sold on the spot market.[28] Still, neither the 1982 pact nor Prime Minister Hussein Musavi's early 1985 visit to Managua generated oil shipments to the hard-pressed Central American state—even though a trade agreement was announced in January 1985 (to exchange Iranian oil for Nicaraguan sugar and meat), and White House spokesman Robert Sims asserted that Teheran was furnishing petroleum and weapons.[29] In Mexico City, Musavi claimed that "for several years, Nicaragua has been purchasing oil from the Islamic Republic of Iran. Like many other countries there's nothing new in this regard this year."[30] An analysis of actual deliveries to Puerto Sandino contradicts his assertion.

A deal involving 9,000 bpd of Ecuadorean crude destined for Managua's Esso refinery collapsed in March 1985. A spokesman for Petraco Ltd. of Bermuda, the intermediary in the aborted transaction, stated that loading of the first cargo was discontinued "upon orders by the Ecuadorean government" to Flopec, the state shipping company.[31] Officials of Petraco and the Ecuadorean government took turns blaming the other for the collapse of an arrangement that reportedly involved 1.2 million barrels of crude in a multilateral swap. Evidently, the venture fell through because of opposition from both Ecuador's conservative armed forces and President León Febres Cordero, who, nonetheless, contended as late as April 1985 that he was prepared to consider any commercially attractive offer that Comandante Ortega might make. This was a gratuitous invitation in light of Nicaragua's acute shortage of foreign exchange as it ran a $500 million annual current account deficit. Similarly, talk of obtaining crude from Peru or Algeria had not produced a substantive agreement as of January 1, 1988, although Peru supplied 34,000 barrels of fuel oil in September 1987.

Fidel Castro was more forthcoming. In early December 1984, Petronic Director Otto Shaffer reported that Cuba would ship Nicaragua 14,285 barrels of liquefied natural gas in December 1984.[32] He did not say whether the cargo would be a donation or sale. Four months later, the Havana regime dispatched a tanker carrying 74,436 barrels of diesel fuel to Corinto.[33] Again, the terms of the transaction were not disclosed, but Managua was bereft of hard currency.

Fundamentally, it was the Soviet Union that filled the gap left by diminishing Mexican shipments. It dispatched tankers laden with Urals grade crude from the Black Sea port of Odessa to Puerto Sandino. Ortega, who had characterized his relations with the Soviet Union as "exemplary," observed that when the United States cut his credit lines to purchase wheat and oil, the USSR was the first country to react by sending wheat and oil to Nicaragua. "We have found in the Soviet Union an attitude of friendship, of respect, of solidarity," he said.[34] As seen in table 17, the USSR moved from a supplemental source to become Nicaragua's dominant supplier, providing almost 90 percent of the Sandinistas' needs in 1985 and—in concert with its allies—nearly 100 percent of its oil imports in 1986 and 1987.

Oil figured prominently on the agenda of Ortega's visit to the Soviet Union, included in a twenty-five-day trip to Cuba and thirteen nations in Eastern and Western Europe during the spring of 1985. Following receipt of the "No cash—no oil" message from de la Madrid on April 16, Ruiz telephoned Managua to suggest that Ortega do something to ensure their country's oil supply. The next day Ruiz informed Mexican officials that the Nicaraguan chief executive would ask Moscow to pick up the slack left in the wake of Mexico's tough new policy.[35] After Ortega's April 29 meeting with Soviet Premier Mikhail Gorbachev, Tass reported that Moscow "will continue to give friendly Nicaragua assistance in resolving urgent problems of economic development, and also political and diplomatic support in its efforts to uphold its sovereignty."[36] However, no mention was made of the nature of any new aid, although a harbinger of expanded Soviet assistance came in the establishment of a joint commission on cooperation in economics, trade, and science and

TABLE 17

Oil deliveries to Nicaragua, 1981–1987

(av. bpd)

	From Mexico	% of Total	From Venezuela	% of Total	From Soviet Union	% of Total	From Other Sources	% of Total	Av. Daily Consumption
1981[1]	5,575	44.7	6,898	55.3	0	0.0	0	0.0	12,473
1982[1]	7,100	55.5	5,700	44.5	0	0.0	0	0.0	12,800
1983[1]	11,300	94.3	0	0.0	685	5.7	0	0.0	11,985
1984[1]	5,800	48.5	0	0.0	5,357	44.8	810[2]	6.8	11,967
1985	1,370	10.4	0	0.0	11,615	88.1	204[3]	1.5	13,189
1986	137	.9	0	0.0	14,658	99.1	0	0.0	14,795
1987	137	1.9	0	0.0	12,811[4]	98.2	93.2[5]	.7	13,071[6]

Sources: Petróleos Mexicanos; Petróleos de Venezuela, S.A.; Petronic; *Latin American Weekly Report*; *Washington Post*; *Daily Report (Latin America)*.

1. Figures cover August through July of the following year for Mexico and Venezuela.

2. Aviation fuel imported from the United States via Aruba and Panama.

3. Shipment of 74,436 barrels of diesel from Cuba landed in Corinto aboard the oil tanker *Cielo di Napoli*, according to the FBIS *Daily Report (Latin America)*, March 11, 1985, p. 12.

4. Of the 12,811 barrels of oil supplied by the Eastern bloc as of September 1, 1987, the Soviet Union provided 59.9 percent, followed by Cuba (15 percent), East Germany (13.5 percent), Bulgaria (5.7 percent), Poland (4.5 percent), and Hungary (1.5 percent). In early September, Ortega was seeking an additional 385,000 barrels from Moscow's satellites; see *Daily Report (Latin America)*, September 8, 1987, p. 7.

5. Nicaragua drew against a $10 million line of credit to purchase 34,000 barrels of oil, delivered by the Peruvian naval tanker *Zorritos* on September 11, 1987; see *Platt's Oilgram News*, September 9, 1987, p. 2.

6. Average daily consumption for the entire year based on oil imported January–October; the Sandinistas were seeking another 700,000 barrels of oil at year's end.

technology. No details were released, but the Russian press described the accord as a "new step on the way of a further development and deepening of mutually beneficial and equal trade and economic relations."[37] Upon returning to Managua three weeks later, Ortega stated that the Kremlin had agreed to supply 80 to 90 percent of Nicaragua's petroleum needs in 1985. Such help was particularly impressive because the Soviets, desperate to sell their oil and gas for dollars on the open market, had been increasingly Scrooge-like in furnishing energy assistance to their satellite states. Of course, Ortega emphasized that receipt of Russian oil would not signify Nicaragua's alignment with the Soviet Union—a charge that had been leveled, he said, "to shock the American people." "We have found in the Soviet Union an attitude of friendship, of respect, of solidarity," Ortega added. "We would like the same relationship with the United States that we have with the Soviet Union."[38]

Ortega's visit to Moscow and the announcement of a top-level formal commitment by the Kremlin to provide oil proved expensive in public opinion in the region, as well as in Washington where the Reagan administration used the pact to mobilize congressional support for $27 million in aid to the contras. Even de la Madrid publicly criticized Managua's revolutionaries. In a June 1985 interview with the Madrid newspaper *El País*, he said: "The Sandinistas should carry out their promises to create a pluralistic democracy and to respect liberties as well as to be nonaligned with the Soviet bloc."[39] Despite these costs, Nicaraguan officials insisted that Ortega had no alternative. "It was a strategic matter. We simply didn't have oil," a government source reported. "He had to conclude an agreement very formally and at the highest levels to avoid the risk of a U.S. naval blockade that would block the supply of oil. We knew there was a price to it."[40] U.S. officials estimated that Soviet Bloc aid to Nicaragua in 1986 totaled $835 million, $300 million of which was military assistance.[41]

As encouraging as the USSR's commitment appeared initially, it was not without negative aspects. Delivery of an odd batch of 16,000 barrels of aviation fuel in 1984 indicated that Soviet oil does not always meet industry specifications. And if Soviet and Cuban advisers should pour into the energy sector along with their oil—something that had not occurred as of early 1988—Esso, whose parent company is under pressure from anti-Sandinista activists in the United States,[42] could pull out, concluding that its profits could never be repatriated.[43] Such a move would leave the inadequately staffed Petronic to operate an aging refinery already bedeviled by heavy maintenance demands and a high turnover rate among its sixty to seventy employees, especially professionals, who seek opportunities outside the country. While no strikes have disrupted output, the government's drafting of young workers adversely affected refining activities until early 1986 when the government apparently realized the problems caused by its conscription policy. Intense Soviet or Cuban influence would probably impel the exodus of Texaco, Chevron, and Shell, which run service stations throughout the country in addition to the sixty operated by Esso.

Officials of Nicaragua's State Energy Institute have insisted that refinery operations would continue despite the U.S. embargo—with replacement parts obtained in Europe, Argentina, Brazil, or other markets. In an interview, Cordero Cuadra of the Nicaraguan embassy raised the Cabinda analogy as a possible reason for the multinational firms' remaining in his country.[44] However, the difference is that Chevron/Gulf Oil turns a handsome profit in Angola, while Esso's ability to repatriate dollars is severely limited. As a result, the firm has insisted that the dollar-starved central bank provide it with dollars in order to purchase spare parts for the refinery.

Stepped-up Russian petroleum activity could also precipitate renewed attacks by anti-Sandinista contra forces against storage depots, pipelines, and vessels. Nicaragua earmarks about half of its oil for military activities, and, in late 1986, stocked a thirty-day petroleum reserve at the Puerto Sandino harbor.[45] This depository would be a natural target for the insurgents. Above all, growing reliance on Soviet supplies enabled U.S. critics of the *comandantes* to justify the economic embargo and to urge officials in Washington to keep military pressure on the revolutionary regime.

Possibly to improve the climate for an arms control agreement with the United States, the Kremlin reportedly decided to curtail petroleum shipments to Nicaragua in 1987. In early June 1987, Ruiz announced that the Soviets would be able to meet only 40 percent of the country's needs for the year "because this socialist country has difficulties with its natural reserves at the present time."[46] Still, the move seemed to be more symbolic than substantive, for Moscow strongly encouraged Cuba, East Germany, Bulgaria, Hungary, Poland, and Czechoslovakia to allocate a portion of their Russian-supplied oil to Nicaragua—an act of socialist solidarity in which the Eastern bloc nations begrudgingly participated. Hence, Urals oil continued to arrive at Puerto Sandino, but with a portion of the cost shifted from Moscow to a satellite.[47] In addition to sending a signal to Washington, the Soviet leaders may have wanted to encourage greater efficiency and conservation in Nicaragua's energy sector where, despite a tripling of prices in mid-1987, gasoline sold for the equivalent of eleven cents per gallon and waste and inefficiency abounded. A detractor of the regime imputed Machiavellian motives to Ruiz in publicizing Russia's decision. "This is a political game with two sides," he stated. "Inside the country, it helps divert people's attention from the real reason for the acute shortage we are feeling right now, which is Sandinista incompetence. Externally, it lays the groundwork for demanding oil credits as an economic concession in exchange for Nicaraguan acceptance of Costa Rican President [Oscar] Arias's peace plan at the summit at the end of June."[48] Nevertheless, Latin American countries, led by Mexico and Argentina, saw the change in Soviet policy as an opportunity to "recover" Nicaragua and disprove the Reagan administration's argument that the Sandinista revolution constituted a communist beachhead on the American mainland. Yet, Caracas's opposition, Mexico City's fixation on the presidential transition, and Buenos Aires's economic distress combined to shelve a scheme, under discussion in mid-1987, whereby non–oil exporting nations such as Argentina would help finance shipments to Nicaragua by Mexico and, possibly, Venezuela.[49]

Just as López Portillo's relations with the Sandinistas in the 1970s epitomized his country's pursuit of regional leadership, de la Madrid's Nicaraguan policy in the mid-1980s characterized Mexico's transition to a Responsible Debtor. Economic considerations at home and the imperative to gain support from Washington and international financial sources militated against the Third World activism of López Portillo, Sepúlveda, Navarrete, Labastida, and others who believed that generous infusions of oil, cash, and technical assistance would promote pluralism, civil liberties, and tolerance within Nicaragua. Mexican pragmatism increased in inverse proportion to its shipments of petroleum to the Managua regime. One official captured the new, pragmatic mood in Mexico City when he said: "Selling oil is a business, and the Nicaraguans haven't been able to pay."[50] Even Mexico's willingness in late 1986 and early 1987 to make modest exports of crude oil to the Sandinistas was contingent upon compensation for this black gold.

The Emergence of
a Contentious Partner?

Mexico's assuming the role of a Responsible Debtor may prove a temporary phenomenon. Any number of conditions could trigger a return to an ideologically motivated foreign policy. These include plummeting oil prices (or even a substantial rise in oil prices), deteriorating economic conditions and an inability to control inflation, international intractability on future debt negotiations, and villification of Mexico in a U.S.-originated campaign against narcotics.

Obviously, Salinas de Gortari will give direction to Mexico's international course. Attempting to predict policies of presidents-designate is more of an art form than exact science in a political system often characterized by pendular changes.[1] Incumbents, who ultimately select their successors, have frequently tried to project their influence into the next *sexenio* by choosing individuals whose views were deemed compatible with their own. Apparently, Adolfo Ruiz Cortines, who named Adolfo López Mateos in 1954, obtained the continuity that he desired. In contrast, conservative Díaz Ordaz was chagrined by the shrill populism and impetuous internationalism that highlighted Echeverría's term. Several elements—the deemphasis on ideology within the PRI, the imperative to possess skills in order to manage the huge public sector, and the salience of economic issues—now work in favor of continuity over change as the mantle of authority passes from one enormously powerful leader to another.

In 1987 the three leading precandidates for the presidency, Energy Secretary Alfredo del Mazo, Interior Secretary Manuel Bartlett Díaz, and Planning and Budget Secretary Salinas de Gortari had distinguished themselves as loyal proponents of de la Madrid's pragmatism in international affairs, which was fashioned—in part—to promote a new development model keyed on structural reforms and non-petroleum exports.

Ultimately, de la Madrid selected Salinas, despite the fact that the thirty-nine-year-old SPP head was the dark horse among journalists, academics, and the roughly 500 members of the *cúpola*, as the upper echelon of Mexico's revolutionary family is known. A number of factors worked in Salinas's favor: his emphatic loyalty to the president, a fellow alumnus of Harvard University;

his political ties to the nation's north (Salinas's father represents Nuevo Leon in the Mexican senate), a region that is most antagonistic to Mexico City in general and to PRI's authoritarian rule in particular; his having assembled a dynamic brain trust, headed by Deputy Planning and Budget Secretary Pedro Aspe Armella, who succeeded Salinas as secretary; his familiarity with both world-scale economic questions and the leaders of international financial institutions who wrestle daily with questions of debt, protectionism, and commodity prices; his undisputed intelligence, which enabled him both to master the intricacies of the economy and to direct economic policymaking following the resignation of Silva Herzog, whom Salinas outmaneuvered in intracabinet budget disputes; his austere, low-key comportment during the selection process compared to the petulant del Mazo's extravagance and flamboyance and Bartlett's overt attempt to broaden his political base and demonstrate proficiency in economic matters; and an unswerving commitment to liberalization, export restructuring, and the tumbling of Mexico's towering trade barriers. As political scientist Luis Rubio F. observed: "Salinas, who was inconceivable as a candidate before 1985, became the overwhelmingly logical choice by 1987."

Mindful of the negligible influence that he will exert on policy after relinquishing the presidential sash, de la Madrid has taken steps to promote continuity between his administration and the next. Second only to picking Salinas, his nation's membership in GATT constitutes the most formidable bridge to the future. Consonant with becoming the organization's ninety-second member in August 1986, Mexico (1) joined the GATT Codes on Standards, Licensing, Customs Valuation, and Subsidies, (2) exempted 90 percent of its tariff items from import permits, (3) bound tariffs on 373 products, and (4) lowered its maximum tariff rate to 20 percent in late 1987, ten months before a decrease to 30 percent was promised. As part of integrating his nation into the international economic system, de la Madrid signed a 1985 Bilateral Subsidies Understanding with the United States, stipulating that his country would phase out export financing by early 1987 in return for receiving an injury test in countervailing duty litigation. To build on this important foundation, the two nations entered into a Bilateral Trade and Investment Framework Agreement in early November 1987. While such an accord appears to be more symbolic than substantive at first, it creates a consultative mechanism for resolving disputes in two often nettlesome areas. Essential to Mexico's economic opening to the world is Japan's decision to invest up to $1 billion during the final eighteen months of de la Madrid's sexennium. At the heart of the commitment is the Petroleum Project of the Pacific to facilitate the transport of crude oil from southeastern production areas to Pacific coast ports whence it could be shipped to Japan, which imports 180,000 bpd, and to other Far East buyers. Additional ventures include $250 million to improve the country's oil infrastructure and

$500 million to support both Mexico's export promotion program and the development of the Sicartsa II steel complex, also on the Pacific coast.[2]

Even if he wanted to (and all the evidence is to the contrary) Salinas would have a hard time undoing the de la Madrid–impelled structural reforms without disrupting the economy. He will also find it difficult to jettison the policy of realistic exchange rates, which—along with weak domestic demand—has been critical to export diversification. Nonpetroleum exports shot up 28 percent during the first half of 1987 over the previous year as Mexico registered a $4.8 billion trade surplus, which—combined with the capital inflows before the October crash of the Mexican stock market—boosted international reserves to a record $15 billion.

Although structural reforms accomplished by a Responsible Debtor may not show conspicuous results until the 1990s, in terms of reducing unemployment, curbing inflation, and impelling development, they offer an emphatically better option for Mexico's resuming sustained growth than does the statist, confrontational approach advocated by nationalist-populists. The agenda favored by nationalist-populists—robust protectionism, enhanced import substitution, enormous subsidies, widespread price controls, and an ever bigger and more dominant public sector—has been discredited. Indeed, pursuit of such a program in the 1970s and early 1980s left the Mexican economy prostrate: a modern-day Gulliver bound by red tape at the hands of a surging army of bureaucratic Lilliputians. The nationalist-populists' weakness was revealed by their inability to prevent Mexico from entering the subsidies pact with the United States, joining the GATT, or participating in the 1987 debt restructuring.

Other factors that are encouraging Mexico's continuance as a Responsible Debtor are the severe consequences of unilateral action on the debt question. The May 1987 action by Citicorp to increase its loss reserves by 150 percent to $5 billion and, hence, to gain greater flexibility in managing its burden of nonperforming loans to Third World states may make it more difficult for Mexico to obtain another debt restructuring when the 1987 plan expires (which will occur shortly before de la Madrid leaves office).[3] Nevertheless, foreign governments and financial institutions have been extremely attentive to Mexico's needs as was evidenced by a December 1987 exit-bond plan, which—with U.S. Treasury backing—could allow the country to retire some $14.5 billion in external debt. However, Mexico's acting unilaterally to limit or repudiate debt repayment could set its economy back a decade and enlarge unemployment dramatically in a country where between 800,000 and 1 million young people enter the labor market each year. Among the probable consequences of such a gambit would be an immediate halt to public and private credits for Mexico's government and businesses, disinvestment, a reduced inflow of capital and renewed capital flight, cash-and-carry trade operations, the exodus of more undocumented workers, worldwide adverse publicity, and intense pressure by U.S. governmental and financial entities for the country to meet its financial obligations. The economic bitterness, which would be concentrated along the border, could even result in creditors seizing Pemex cargoes or Aeromexico aircraft. Because of the interrelationship between the United States and Mexico, the effects of any freelance debt initiative by the latter would be much more far-reaching than those afflicting Peru, whose GDP is one-tenth the size of Mexico's. A 9 percent growth rate in 1986 aside, Peru suffered drastically because of President García's decision to limit debt repayments to 10 percent of export earnings. The pain took the form of a shortfall in external credits, distorted internal prices, domestic shortages, commercial interruptions, trade union unrest, and unrepatriated foreign exchange.[4] No doubt because he was unwilling to be closely identified with his Peruvian counterpart, de la Madrid greeted García cordially but without enthusiasm when he visited Mexico in the spring of 1987. After all, articles in specialized periodicals had contrasted Mexico's diligence in dealing with its obligations to the mercurial behavior of Peru, which, on August 15, 1986, was declared ineligible for IMF loans for failure to meet payments. In the language of the *Economist,* Peru had become a "pariah" and the fifth member of the Fund's "rogues' gallery" that also included Vietnam, Guyana, Liberia, and Sudan.[5] Especially striking was a juxtaposition of articles in the *Wall Street Journal:* one featured President García's quixotic attempt to play David against the IMF's Goliath; the other described Mexico's prudent and valiant effort to reach an accommodation with its creditors.[6]

Symbiotic Ties

Should Salinas, as anticipated, continue to deepen the structural reforms begun in the mid-1980s, Mexico's economy will become even more enmeshed in the world economy. Increased involvement with Japan, Western Europe, and Canada will accompany this fusion. However, geography, mutual interests, and inertia will further enhance the symbiotic relationship between the United States and Mexico, whose frontier is crossed more than 300 million times a year by residents of the two countries. As is clear from our discussion, the importance of the United States to Mexico is overwhelming, a fact especially evident in economic matters. In 1986 U.S. firms supplied 70.4 percent of Mexico's imports, while buying 76.6 percent of Mexico's imports. Meanwhile, the "in-bond" assembly industry, concentrated in the vicinity of the border where 10 million people live, generated almost 300,000 jobs in Mexico and added $1.5 billion to the value of manufactured goods destined for the United States. In the energy sector, private U.S. companies and the Department of Energy purchased one-half of the 1.34 million bpd of oil that Pemex shipped abroad in 1987. In addition, U.S. banks, which hold nearly one-third of Mexico's external debt, were the principal sources of commercial credit. Mexican-Americans and Mexicans living in the United States returned some $2 to $3 billion in remittances to their families below the Rio Grande, U.S. entrepreneurs accounted for 63 percent of the $19.5 in direct investment in Mexico, and North Americans constituted nearly 85 percent of the 5 million foreign tourists who visited Mexico in 1987. As I have previously mentioned, three times in recent years—1976, 1982, and 1986—the U.S. government spearheaded efforts to shore up a teetering Mexican economy. Indeed, the integration of the two economies is well advanced in the petroleum, automotive, tourist, feedlot cattle, and speciality fruit and vegetable sectors. Unfortunately, illegal drugs may be another item of mutual dependence.

What is true in the economic realm obtains in other areas of Mexican national life. For example, U.S. cultural influence reveals itself in a stream of motion pictures, television programs, books, news service articles, and magazines. *Selecciones del Reader's Digest* is one of the best-selling magazines in Mexico. Monday night football and major league baseball games, especially those featuring Sonora-born Fernando Valenzuela, are telecast live to Mexico City.[7] English, which long ago replaced French as the country's leading foreign language, has invaded Mexican Spanish as reflected in the widespread use of "okay," "goodbye," "maybe," and scores of other words, and a Mexican Christmas would not be complete without Santa Claus. Half of the persons interviewed in a late 1986 survey conducted by the *New York Times* said that they had close relatives living in the United States, a country about which opinion was quite favorable.[8] In military matters, Mexico, while gradually modernizing its armed forces, boasts one of the world's lowest per-capita expenditures on defense ($14) thanks to Pentagon-ensured protection against external invasion.

While the bilateral relationship is profoundly asymmetrical, Mexico's importance to the United States is also noteworthy. As mentioned above, in 1986 Mexico, which ranked behind Canada, Japan, and West Germany as the United States' fourth leading trading partner, provided a $12.4 billion market for U.S. exports, and shipped $17.6 billion in light manufactures and primary products to the country. Sales to Mexico generated 500,000 jobs for North American workers. All told, Pemex was the largest single contributor to the U.S. Strategic Petroleum Reserve. Mexican-born workers are considered crucial to the success of agriculture in California's Imperial Valley, as well as to assembly, textile, and construction industries throughout the Southwest.

Culturally, Mexican influence is manifest in art, literature, clothing, food, and language. Through Univision, the U.S. branch of Televisa, Mexicooriginated Spanish language telecasts are carried on 300 stations and viewed by an audience of between 6 and 10 million Hispanics. Mario Vázquez Rana's purchase of United Press International should expand the wire service's coverage of events in Mexico. At the popular level, the increase to 2,409 (as of 1987) in the number of Taco Bell restaurants, founded in the early 1960s, attests to the popularity of Mexican-style food in the United States.

Militarily, the presence of a stable neighbor to the south has obviated the need for more than token forces along the Rio Grande, the longest border in the world between a developed and a Third World country. This secure frontier has enabled the United States to project its power in Europe and, to a lesser extent, in Asia without fearing pressure on its southern flank. Mexico, where approximately 500,000 North Americans live, has also served as a buffer to mitigate the spillover into the United States of men, women, and children deracinated by the armed conflict scourging Central America. As a consequence, one public opinion survey found that 97 percent of U.S. leaders and 75 percent of the U.S. public believed Mexico to be of "vital interest" to the United States.[9] Tables 18 through 23 reveal the multiplying, inextricable ties that bind Mexico and the United States in mutual trade, investment, communication, travel, and migration.

TABLE 18
Bilateral Economic Relations: U.S.–Mexican Trade, 1970–1986

	Mexican Exports to U.S. ($ millions)	% of All Mexican Exports	Mexican Imports from U.S. ($ millions)	% of All Mexican Imports
1970	839	65.3	1,568	62.7
1971	911	65.1	1,479	61.6
1972	1,288	75.8	1,745	58.2
1973	1,318	58.8	2,277	54.2
1974	1,703	58.7	3,779	62.2
1975	1,668	53.8	4,113	57.9
1976	2,111	60.9	3,774	62.5
1977	2,738	59.5	3,493	58.2
1978	4,057	68.1	4,564	60.4
1979	6,252	69.6	7,563	62.6
1980	10,072	63.2	11,979	65.6
1981	10,716	55.3	15,398	63.8
1982	11,887	52.0	8,921	59.9
1983	13,034	n.a.	4,958	n.a.
1984	14,612	60.4	6,695	59.3
1985	15,029	68.6	11,132	82.5
1986	17,302	76.6	12,382	70.4

Source: International Monetary Fund, *Direction of Trade Statistics Yearbook* (Washington, D.C.: IMF, 1970–1986); telephone interview with Brent Fogt, Mexico Division: U.S. Department of Commerce, June 3, 1987.

Evolution of Mexico's Political System

Structural reforms and increasing interactions with the United States should foster greater pluralism in Mexico's political system—the former because market mechanisms, once implanted in the economy, will influence political behavior; the latter because Mexican elites are extremely sensitive to the views that North Americans hold of their regime.

Since Echeverría left office in 1976, key Mexican policymakers have attempted to change incrementally the PRI in particular and the political system in general. After all, President Plutarco Elías Calles founded the ruling party when Mexico, then an overwhelmingly rural, agrarian, and parochial nation, lacked educational opportunities for all but the wealthy and a small middle class. Since World War II, an increasingly urban, industrialized, and cosmopolitan Mexico has boasted a well-educated middle class that comprises approximately one-quarter of the population. In response to the pressure for greater citizen participation in decision making, the López Portillo regime expanded freedom of the press, facilitated the registration of new political parties, guaranteed opposition parties 100 seats in an enlarged Chamber of Deputies, and used the Congress as a forum for televised debates on critical issues.[10] De la Madrid advanced the so-called democratization process by

TABLE 19
Bilateral Economic Relations: U.S. Direct Investment in Mexico, 1970–1986
($ millions)

	Investment in Mexico	Investment in All Countries	% in Mexico	Investment in Developing Countries	% in Mexico
1970	1,786	78,178	2.3	21,448	8.3
1971	1,838	86,198	2.1	25,358	7.2
1972	2,025	94,337	2.1	25,235	8.0
1973	2,379	103,675	2.3	25,266	9.4
1974	2,825	118,613	2.4	28,479	9.9
1975	3,200	124,212	2.6	26,222	12.2
1976	2,984	137,224	2.2	29,050	10.3
1977	3,230	149,848	2.2	34,462	9.4
1978	3,690	167,804	2.2	40,399	9.1
1979	4,490	186,750	2.4	44,525	10.1
1980	5,989	215,578	2.8	53,277	11.2
1981	6,977	226,359	3.1	56,182	12.4
1982	5,544	221,512	2.5	52,441	10.6
1983	4,999	226,117	2.2	50,978	9.8
1984	4,568	212,994	2.1	50,131	9.1
1985	5,087	232,667	2.2	54,474	9.3
1986	4,826	259,980	1.9	60,609	8.0

Source: U.S. Department of Commerce, Bureau of Economic Analysis, *Survey of Current Business* (Washington, D.C.: Government Printing Office, 1970–1986).

TABLE 20
Bilateral Communications: U.S.–Mexican Telephone and Telegraph Messages, 1970–1985

		Mexico to United States			United States to Mexico		
		No. of Messages	Total Minutes	Cost ($)	No. of Messages	Total Minutes	Cost ($)
Telegraph Messages	1970	11	361	32	37	1,467	83
	1971	334	334	34	50	2,406	123
	1972	1	46	4	17	1,281	75
	1973	1,798	81,152	3,334	12,712	612,735	47,407
	1974	1,824	81,313	3,062	12,228	572,735	47,407
	1975	2,115	84,636	3,177	12,710	657,612	26,368
	1976	4,730	185,814	7,361	14,551	803,462	34,983
	1977	4,146	148,151	8,615	11,114	653,198	36,212
	1978	4,348	149,550	9,332	12,138	737,895	43,187
	1979	4,419	153,407	9,826	16,333	1,025,249	60,693
	1980	4,401	153,799	9,556	16,146	1,046,064	62,387
	1981	4,613	165,175	10,853	27,803	1,073,928	123,094
	1982	3,409	115,588	7,688	25,728	1,938,222	123,856
	1983	2,083	69,083	4,235	18,027	1,454,007	101,131
	1984				20,780	1,397,271	129,601
Telephone Calls (in millions)	1978	10.7	n.a.	69.9	17.0	n.a.	112.2
	1979	14.3	n.a.	93.6	21.6	n.a.	186.3
	1980	18.4	n.a.	126.4	26.1	n.a.	220.7
	1981	22.4	n.a.	173.6	29.8	n.a.	294.7
	1982	25.8	155.6	175.0	36.4	251.8	317.0
	1983	26.7	144.5	146.3	39.3	294.3	334.5
	1984	25.9	187.1	149.4	41.7	307.0	361.4
	1985	30.4	170.0	169.2	45.6	336.1	381.4

Sources: U.S. Federal Communications Commission, Statistics of Communications: Common Carriers (Washington, D.C.: Government Printing Office, 1971–1983); telephone interview with Kenneth Stanley, Industry Analysis Division, Federal Communications Division, June 3, 1987.

holding primary-style *consultas de base* to select PRI nominees for city councils and state legislatures and by creating a Federal District Representative Assembly in Mexico City where the main independent parties of the left—the Unified Socialist party (PSUM) of Mexico, the Mexican Workers' party (PMT), and the Trotskyite Revolutionary Workers' party (PRT)—garnered 57 percent of their nationwide vote in 1985.[11] Another reform gave opposition parties 40 percent of the seats in the Chamber of Deputies, which was enlarged to 500 members.[12] Furthermore, despite discontent expressed by the labor sector and by other traditional elements within its ranks, the PRI made a serious effort to nominate more attractive, locally popular, and better qualified candidates for gubernatorial elections held in 1986, 1987, and 1988.[13]

Did President de la Madrid risk a social upheaval by attempting to implement structural and political changes amid exceedingly harsh conditions? Despite an infusion of external credits and higher oil prices, Mexico was afflicted by acutely difficult circumstances as of late 1987. Consumer prices were climbing at an average annual rate of 141 percent, exciting speculation that the govern-

TABLE 21
Bilateral Communications: Foreign Visitors
to the United States, 1970–1986

	Visitors from All Countries	Visitors from Mexico	% of Total
1970	12,362,299	1,058,772	8.6
1971	12,739,006	1,170,583	9.2
1972	13,057,119	1,377,143	10.5
1973	13,995,164	1,619,451	11.6
1974	14,123,253	1,840,849	13.0
1975	15,698,118	2,155,651	13.8
1976	17,523,239	1,920,509	10.9
1977	18,610,000	2,700,000	14.5
1978	19,842,000	2,100,000	10.6
1979	20,310,000	2,600,000	12.8
1980	22,300,000	3,323,000	15.0
1981	22,100,000	1,767,000	8.0
1982	18,600,000	1,548,000	8.3
1983	21,700,000	669,000	3.1
1984	20,800,000	823,000	4.0
1985	19,298,000	2,731,000	14.2
1986	22,003,000	3,900,000	17.7

Sources: U.S. Department of Commerce. Office of Research and Analysis, U.S. Travel Service, *Foreign Visitor Arrivals, 1966–1976* (Washington, D.C., Government Printing Office); telephone interview with Karen Woodrow, U.S. Department of Commerce, Bureau of the Census, June 2, 1987.

Note: Except for Canadian visitors, arrival figures include persons traveling for business, pleasure, study, and transit. Arrival figures exclude Canadian and Mexican border crossings.

ment would freeze wages, prices, and the exchange rate as a shock treatment to halt inflation, and GDP had risen less than 1 percent from the year before. The minimum wage lagged behind inflation for the fifth straight year, as worker purchasing power had sagged 45 percent since de la Madrid took office; open and disguised unemployment beset nearly half of the work force of more than 25 million people; the foreign debt climbed to $108 billion; subsidies were declining even as the price of such essential items as gasoline, electricity, transportations, and food was rising; before and after the October 19, 1987, "Black Monday" on the New York stock exchange, the Mexican stock index plunged as much as 75 percent; a month later the peso fell 25 percent against the dollar as the Central Bank, anxious to conserve its hard currency reserves, withdrew its support for the peso on the secondary or "free" market; and ubiquitous sources of pollution continued to darken the already fetid air of Mexico City and other metropolises. Similar conditions in the United States or Western Europe would have ignited demonstrations, mobilized strikes, and launched the careers of demagogic politicians. Nevertheless, Mexico remained remarkably quiet—with discontent manifest largely in (1) *panista* cries of political fraud, (2) strident speeches by union spokesmen, opposition party leaders, and intellectuals, (3) bristling editorials in *Unomásuno, La Jornada,* and other leftist newspapers, (4) muckraking articles in *Proceso* about antinationalistic and corrupt acts by government officials, (5) criticism of the regime by leaders of the country's Roman Catholic church, (6) an upswing in petty crime in Mexico City, (7) acute cynicism and pessimism registered in public opinion surveys, and mounting voter abstentionism, and (8) demands for reform of the PRI by the self-described Democratic Current, one of whose

TABLE 22
Bilateral Economic Relations: U.S.– Mexican Travelers' Expenditures, 1970–1985
($ millions)

	Mexican Travelers to the United States	U.S. Travelers to Mexico		Mexican Travelers to the United States	U.S. Travelers to Mexico
1970	583	778	1978	1,456	2,121
1971	565	959	1979	1,975	2,460
1972	620	1,135	1980	2,522	2,564
1973	830	1,264	1981	3,775	2,862
1974	1,142	1,475	1982	3,098	3,324
1975	1,490	1,637	1983	1,951	3,618
1976	1,364	1,723	1984	1,899	3,609
1977	1,316	1,918	1985	2,130	3,552

Source: U.S. Department of Commerce, Social and Economic Statistics Administration, Bureau of the Census, *Statistical Abstract of Latin America* (Washington, D.C.: Government Printing Office).

leaders—Cuauhtémoc Cárdenas (son of the late president) bolted the official party to accept the presidential nomination of the small, right-wing Party of the Authentic Revolution.[14]

Several factors explained the relative tranquility. First, the several years of prosperity, when GDP grew at an 8 percent annual rate, that preceded de la Madrid's election gave his administration some breathing space—a "political cushion," in the words of Salinas de Gortari[15]—that the next president will not enjoy. Second, despite a sharp decrease in the buying power of workers, employment did not decline as dramatically as anticipated, according to official figures.[16] Third, the barter or underground economy expanded to the point where it equaled 25 to 35 percent of GDP in the mid-1980s. Although they deprived the national treasury of resources equivalent to 26 percent of the revenues, transactions in this *economía subterránea* helped mitigate the difficulties faced by millions of Mexicans.[17] Fourth, extended families, which often have several members working full-time and others selling artificial flowers, washing windshields to obtain handouts, or blowing kerosene-induced flames from their mouths in city traffic jams, provided a safety net for upwards of one-third of the work force in a country where social security coverage is limited and unemployment compensation nonexistent. Fifth, for Mexicans, the most attractive welfare scheme was a porous U.S. frontier across which nearly

TABLE 23
Legal and Illegal Mexican Migration to the United States, 1970–1986

Fiscal Year	No. of Legal Immigrants	Illegal Aliens Apprehended	Total Apprehensions	% Mexicans
1970	44,469	277,377	345,353	80.3
1971	50,103	348,178	420,126	82.9
1972	64,040	430,213	505,949	95.0
1973	70,141	576,823	655,968	87.9
1974	71,586	709,959	788,145	90.1
1975	62,205	680,392	766,600	88.8
1976	57,863	781,474	875,915	89.7
1977	44,079	954,778	1,042,215	91.6
1978	92,400	976,667	1,057,977	92.3
1979	52,100	988,830	1,076,418	91.9
1980	56,680	817,381	910,361	89.8
1981	101,268	874,161	975,780	89.6
1982	56,106	887,457	970,246	91.5
1983	59,079	1,172,306	1,251,357	94.0
1984	57,557	1,170,769	1,246,981	93.9
1985	61,077	1,348,749	1,348,749	94.0
1986	66,533	1,767,400	1,767,400	95.0

Sources: U.S. Department of Justice, Immigration and Naturalization Service, *1984 Statistical Yearbook* (Washington, D.C.: Government Printing Office); telephone interview with Karen Woodrow, U.S. Department of Commerce, Bureau of Census, June 2, 1987.

2 million illegal immigrants passed in 1986. In December 1986, President Reagan signed an immigration reform bill, but its provisions are so flexible—notably with respect to agricultural workers—that the legislation has only slowed, not halted the heavy influx of unlawful workers.

Sixth, even though the specter of a growing PAN perturbs the PRI, political cannibalism has engulfed the left, where disputatious parties have sprouted like mushrooms in a dank cave. Over the years, hardliners have competed with Eurocommunists for leadership of the PSUM, a left-wing amalgam that captured only 3.2 percent of the vote in the 1985 congressional elections. It remains to be seen whether the PSUM and four other leftist parties, which joined together to form the Mexican Socialist party (PMS) in 1987, can work in harness behind a single candidate, Herberto Castillo, in the upcoming presidential campaign.[18] Neither the PRT nor the government-subsidized Popular Socialist party (PPS) and Workers' Socialist party (PST) entered the new grouping, which, as of late 1987, had not expanded its base beyond intellectual, academic, and marginal labor elements. Cárdenas will attract votes from the left as will Rosario Ibarra de la Piedra, a respected human rights activist who has accepted the nomination of the Trotskyite PRT. As far as the PRI is concerned, the more candidates the merrier. Thus far, knowledge both that violence accomplishes nothing politically and that the army, which massacred hundreds of middle-class protesters in 1968, will brook no guerrilla adventurism has kept extremists in PMS, the PRT, and the Castro-oriented *Corriente Socialista* from trading ballots for bullets. An exception has been the minuscule Mexican Vicente Guerrero Revolutionary Command, which in May 1987 claimed credit for an explosion in front of the PRI's Mexico City headquarters that shattered windows and destroyed an ambulance owned by the party.[19] The death of more than one million people in the revolution that erupted in 1910—known to adults who learned of it from grandparents who vividly remembered the carnage—has sensitized Mexicans to the danger inherent in widespread violence. The 1968 Tlatelolco massacre provided a more recent reminder of the regime's willingness to use force in a confrontation.

Seventh, the 145,000-man armed forces, though top heavy with senior officers and hardly of Prussian caliber, has started to transform themselves into a modern military. Jeeps and trucks have replaced horses in all but one of twenty-eight cavalry units; five armored regiments boast personnel carriers and light Panhard tanks; German-designed G-3 automatic rifles have been introduced in the army's eighty-five infantry battalions; and the air force, a branch of the army, has acquired a dozen F-5 supersonic aircraft. Military leaders continue to trumpet their loyalty to a political system that has co-opted them with a lavish array of social and economic perquisites, including attractive salary increases for key officers during the 1981–1988 period. To enhance efficiency, responsiveness, and accountability, de la Madrid approved plans to collapse the army's thirty-six military zones into several corps, the first of which will be in the Mexico City area. Even more worrisome than an organized challenge to the system is the remote possibility that a spontaneous demonstration in one part of Mexico City, say, a protest over higher tortilla prices or bus fares, could spread like wildfire across the capital and, from there, to other cities, unimpeded by the firebreak of legitimacy that is essential to any regime's survival. Still, a loyal military, combined with government control over the mass media, should curb the outbreak of anomic disturbances. Arguably, "too many Mexicans have a stake in social tranquility and political continuity for a politician or faction to find fertile ground for rebellion, pronuciamentos, or radical breaks with the status quo."[20]

Finally, and perhaps most important, Fidel Velázquez has coupled strident advocacy of higher compensation for trade unionists with a readiness to beat a tactical retreat if it appeared that his demands would spur bankruptcies or wreck the administration's structural reform program. Other confederations—the CROC, CROM, CRT, and FSTSE, for example—have generally followed the crusty octogenarian's lead on labor questions. In early 1987, Velázquez acquiesced in the government's decisive, nonviolent quelling of strikes by telephone and electrical workers. If successful, these unions could have crippled vital economic sectors and paved the way for a contagion of labor stoppages by ever more disgruntled workers. De la Madrid and his successor can attempt to propitiate the CTM by lauding its "responsible and nationalistic" leaders, seeking the advice of "Don Fidel" (as he is reverently known) on the selection of PRI nominees for governorships and seats in Congress where labor feels grossly underrepresented, and guaranteeing that the union movement doesn't consistently lose ground to better-educated, middle-class candidates in the official party's grassroots *consultas.* Velázquez, a staunch del Mazo backer, was upset at de la Madrid's naming yet another technocrat as the PRI's candidate. Still, in recent years, the PRI has awarded the CTM more congressional seats to compensate for fewer economic benefits—a strategy likely to be complemented by Salinas's finding attractive bureaucratic posts for more labor loyalists. Economic rewards to labor could replace political payoffs if and when sustained growth occurs. Even though its role is pivotal to the success of a new development strategy, a combination of factors suggest that the CTM's influence will wane. To begin with, Velázquez will pass from the scene, and, while the succession to a new secretary-general should be orderly, no future labor leader will boast his revolutionary credentials or wield his political power. Moreover, the Mexican economy is changing in a manner that finds unionized workers declining as a percentage of a rapidly expanding labor force. The growth in employment is concentrated in *maquiladoras* and other service industries whose employees, like their U.S. counterparts, are notoriously difficult to organize. Above all, stronger emphasis on both market forces and combating corruption militate against the sweetheart arrangements that many unions—the oil workers, for example—have enjoyed with their employers. (See table 24.)

It would be premature to write organized labor's epitaph. While many *charros*, as old-line union chiefs are known, owe their privileges to the government, Salinas's announced goals of economic restructuring and modernization threaten the aging leaders' preeminent positions and their members' pocketbooks. How long will the rank and file, many of whom are one-third the age of their *charros*, remain tranquil if CTM leaders acquiesce in a liberalization program that entails even more sacrifices? La Quina wasted no time in showing his defiance of the president-designate. Meanwhile, a younger breed of activists—Francisco Hernández Juárez of the Telephone Workers, Jorge Tapia Sandoval of the Mexican Electricians Union, and Senator Arturo Romo Gutiérrez of Zacatecas, who heads the CTM's education department—are disposed to confront the regime on economic issues rather than witness a further decline in labor's economic and political strength.

Labor is only the most coherent among the antiliberalization groups. Also hostile to structural reforms are inefficient manufacturers, who fear competition; bureaucrats, who fear that reducing the state's economic role will diminish the influence and opportunities derived from dispensing import permits, fixing quotas, and performing thousands of other regulatory functions; many PRI stalwarts, who fear that expanding the private sector at the expense of the state will shrink the patronage and boodle needed to lubricate the party's creaky machinery; and intellectuals, who fear that encouraging market forces will enhance the role of despised domestic and foreign capitalists. No greater challenge confronts Salinas than that of broadening a coalition—one that embraces large, efficient producers, the international financial community, and many high-level *técnicos* in ministries with economic functions—to advance the reforms for which de la Madrid laid the groundwork. The internal impact of those reforms had barely been felt in early 1988, and their future would be imperiled if an economic slowdown or protectionism were to hamper Mexican exports to the United States.

Contentious Partner

Greater reform-inspired pluralism below the Rio Grande may diminish "Mexico-bashing" in the United States; nonetheless, increased interaction will impose a strain on relations between the two countries. The tension will be all the more acute because of the asymmetry of the growing interdependence.

TABLE 24
Labor's Representation in the Mexican Chamber of Deputies, 1964–1988

Legislature	PRI	Workers' Sector	CTM	Workers' Sector PRI	$\frac{CTM}{PRI}$	CTM Workers' Sector
XLVI (1964–1967)	177	27	17	15.3	9.6	63.0
XLVII (1967–1970)	172	34	14	19.8	8.1	41.2
XLVIII (1970–1973)	150	25	14	16.7	9.3	56.0
XLIX (1973–1976)	173	27	20	15.6	11.6	74.1
L (1976–1979)	262	39	23	14.9	8.8	59.0
LI (1979–1982)	294	66	43	22.4	14.6	65.2
LII (1982–1985)	300	73	50	24.3	16.7	68.5
LIII (1985–1988)	289	71	44	24.6	15.2	62.0

Sources: Ricardo de la Peña and César Zaqueta, *La estructura del congreso del trabajo, estado, trabajo y capital en méxico: un acercamiento al tema* (Mexico City: Fondo de Cultura Económica, 1984); Republic of Mexico, Cámara de Diputados, *Directorio: 1982–1985* (Mexico City: Congreso de la Unión, n.d.); data supplied by Deputy Lic. Fernando Ortiz Arana, Secretary of Electoral Action, PRI.

Compared to its Spanish-speaking neighbor, the United States boasts a population three times larger, a geographic area five times greater, a per-capita income six times higher, and a military seventeen times bigger. As Abraham F. Lowenthal has pointed out, the United States manufactures as many automobiles in a day as Mexico does in a month, while Mexico's entire national product approximates that generated within a sixty-mile radius of downtown Los Angeles.[21]

A sense of vulnerability and fear of losing its sovereignty will make Mexico a Contentious Partner as the two economies become inextricably bound. Symbolic and substantive features will define this new role. In the theater, actors attempt to vivify their characters by commanding selected means, that is, dramatic agents. Shakespeare endowed his characters with guile (Richard III), daggers (Julius Caesar, Hamlet), potions (Romeo and Juliet), and other artifices.[22] Mexican chief executives have their own stock of dramatic conceits, not the least of which is the exchange of visits with communist leaders. To deflect domestic attention from its proliferating ties with the United States, Mexico's president may bestow *abrazos* on Castro, the Ortega brothers, and other well-known leftists. Token oil shipments to the Sandinistas offers another possibility. It was politically useful that, following the renegotiation of Mexico's debt in 1986, Soviet Foreign Minister Eduard Shevardnadze paid a three-day visit to Mexico City. After reciprocating the visit in May 1987, Foreign Secretary Sepúlveda announced that General Secretary Mikhail Gorbachev would fly to Mexico to meet with de la Madrid as part of a Latin American tour. Assurances of trade, technology exchanges, and cultural ties invariably surround such tête-à-têtes; still, Mexican-Soviet economic links remain anemic, and Mexican leaders—publicly and in private conversations with U.S. officials—have stressed their opposition to allowing the Russians, who have only a single consulate outside the capital (in Veracruz), to open additional consulates in Mexican cities along the U.S. border. Gorbachev-type visits inspire U.S. politicians to deliver themselves of "Mexico-is-going-Marxist" rhetoric, which reinforces the image of Mexico's independence of Uncle Sam. To the extent that Mexico becomes a more competitive exporter, the country will be able to diversify its sales abroad.

Mexico's army embodies the nation's sovereignty. For this reason, playing the role of a Contentious Partner militates against significantly expanded contacts between the Mexican and U.S. armed forces, although the upwardly mobile officers invited to attend the National Defense College annually visit the United States. Neither the training of Mexican units in U.S. military facilities nor the conduct of joint maneuvers is likely to take place in the foreseeable future. This situation will not prevent the Mexican navy's continued cooperation with the U.S. Coast Guard in search and rescue operations in the Gulf of Mexico and the Pacific Ocean.

Mexico's sensitivity to the integrity of its northern border will intensify. Consequently, the Mexican government may be even more reluctant than in the past to permit U.S. Customs Service aircraft to undertake hot pursuit deep into Mexican territory as they chase the aircraft of suspected drug smugglers. Still, when the problem is essentially technical and scientific, collaboration will take place, just as it did in 1987 when U.S. and Mexican scientists fought shoulder-to-shoulder to prevent Africanized honeybees or "killer bees" from swarming north of the 125-mile-wide Isthmus of Tehuantepec.[23] The border-focused *maquiladora* industry epitomizes the interdependence of the two countries. On the one hand, cooperation engenders mutual benefits; on the other hand, an adverse U.S. tariff policy would devastate Mexican exports from this sector, while Mexico's nationalization of these plants would imperil North American investment.

As a Contentious Partner, Mexico can be expected to continue its efforts as an Honest Broker, while renewing its quest for regional leadership. One difference could contrast its resurgence as a Regional Leader in the late 1980s and early 1990s to its role during the Echeverría and López Portillo years. Specifically, conversion to an export-driven development model and the harmonization of its economy with those of the First World should cool Mexico's enthusiasm for a new international economic order even as SRE continues to give rhetorical support for a NIEO, just as de la Madrid did when the Group of Eight major Latin American countries met in Acapulco in late 1987.

While Mexico may participate actively in GATT and specialized financial institutions of the United Nations, it will continue to interest itself in the problems of Central America and the Caribbean. Initiatives on peace, disarmament, and human rights will highlight the country's brokership efforts in this region. In view of burgeoning domestic opposition to support for the Sandinistas and the failure of the Contadora process to bring about peace, one scholar observed, "Mexico has literally no place left to go in Central America, yet has nowhere to go in international affairs other than Central America."[24]

Managing a contentious partnership will require extraordinary adroitness in Mexico City and in Washington. No matter how prickly bilateral relations become, the growing mutuality of interests raises the stakes and heightens the imperative to resolve conflicts as efficiently as possible. As I noted in chapter 4, a more effective lobbying effort in Washington would advance Mexico's cause in U.S. legislative, executive, and judicial arenas. Even with an activist embassy—which, ironically, might permit the U.S. embassy in Mexico City to be less assertive—Mexico should work with the United States both to establish clear priorities under the direction of a responsible, senior official and then to keep salient bilateral issues in separate negotiating channels.

The experience of both the Bilateral Energy Consultative Group and negotiators of the trade and investment framework demonstrates the advantages of a segmented over a linkage approach in which energy, trade, migration, narcotics, investments, and other questions might be negotiated together— with tradeoffs attempted between the various items. As attractive as a package deal between the two countries might seem, keeping each issue in its own channel will prove more effective in handling an increasingly complex bilateral agenda. To begin with, compartmentalization allows negotiations to proceed at the speed appropriate to the issue at hand with a relatively small group of participants. As the number of issues increases arithmetically, the cast of participants seems to grow geometrically—with escalating prospects for dis- agreement and conflict. Segmentation also permits discussions to take place among mid-level officials, who are more likely than their superiors to have time to master the issue and to build confidence on a personal basis with their counterparts from across the border. Confining parlays to specialists may have the benefit of reducing both the political content and press coverage of the conversations. Hyperbole about a "special relationship" or "North American Accord" serves only to raise the visibility of discussions and alarm Mexicans about their country's vulnerability vis-à-vis the United States. Mexican analysts express profound trepidation about the "impressive set of potential pressure points" available to U.S. officials in dealings with their southern neighbor.[25] Furthermore, encasing each issue in its own channel enhances damage control if talks in one area turn sour, while preventing the accumulation of pressures and expectations that could explode into a cause célèbre. Obvi- ously, the large cast of public and private actors in bilateral relations and the blurring of foreign and domestic policy in both countries prevents hermetically sealing questions from each other. Still, the separate pursuit of each of six or eight issues will lead to greater cooperation than attempting to put the same number of items in one negotiating basket.

The proliferation of contacts between Contentious Partners will spark fric- tion on the surface of bilateral relations even as deeper, stronger currents impel the two nations closer to each other. Ensuring that the tension is creative and not destructive will challenge the skill of leaders both in Mexico City and in Washington. The failure to manage bilateral affairs in an intelligent, mature, and mutually beneficial fashion will threaten the long-range economic well- being and security of citizens in both great nations.

Notes

Bibliography

Index

Notes

Introduction

1. Mario Ramón Beteta, *Report Presented by the Director General of Petróleos Mexicanos* (Mexico City: Petróleos Mexicanos, 1986), p. 2.

2. "Mexico: Crisis of Poverty," *Los Angeles Times* supplement, July 15, 1979, p. 9.

3. Jorge Castañeda, "Revolution and Foreign Policy: Mexico's Experience," *Political Science Quarterly* 78, no. 3 (September 1963): 404.

4. Howard F. Cline, *The United States and Mexico*, rev. ed. (New York: Atheneum, 1965), p. 241.

5. Quoted in *Excelsior*, October 27, 1977, p. 1-A.

6. Robert O. Keohane and Joseph S. Nye, *Power and Interdependence: World Politics in Transition* (Boston and Toronto: Little, Brown, 1977), pp. 38–60.

7. Robert O. Keohane, "Theory of World Politics: Structural Realism and Beyond," in *Neorealism and Its Critics*, ed. Keohane (New York: Columbia University Press, 1986), pp. 158–203.

8. John. C. Wahlke, Heinz Eulau, William Buchanan, and Leroy C. Ferguson, *The Legislative System: Explorations in Legislative Behavior* (New York: John Wiley, 1962).

9. K. J. Holsti, *International Politics: A Framework for Analysis* (Englewood Cliffs, N.J.: Prentice-Hall, 1983), p. 116.

10. Dean Rusk quoting Kennedy, in K. J. Holsti, "National Role Conceptions in the Study of Foreign Policy," *International Studies Quarterly* 14 (1970): 265; "The Inaugural Address of John Fitzgerald Kennedy," *The Kennedy Presidential Press Conferences* (New York: Earl M. Coleman, 1978), p. v.

11. Holsti, "National Role Conceptions in the Study of Foreign Policy," pp. 267–68.

12. Quoted in ibid., pp. 267–68.

13. These and other roles are discussed in ibid., pp. 260–71.

14. M. Delal Baer, "Mexico: Ambivalent Ally," *Washington Quarterly* 10 (Summer 1987): 111.

15. Edwin J. Wellhausen, "The Agriculture of Mexico," *Scientific American* 235 (September 1976): 129.

16. The Mexican constitution provides that the Congress may declare war (upon examining the facts submitted by the executive), admit new states and territories to the Mexican union, create and maintain the armed forces, regulate the diplomatic and consular corps, help define the legal status of foreign nationals in the country, specify terms under which the president may negotiate loans backed by the nation's credit, and enact tariff legislation. For its part, the Senate ratifies treaties, approves appointments, authorizes the dispatch of troops outside the country, and permits the peaceful passage of foreign troops through Mexican territory. See George I. Blanksten, "Foreign

Policy of Mexico," in *Foreign Policy in World Politics*, 2d ed., ed. Roy C. Macridis (Englewood Cliffs, N.J.: Prentice-Hall, 1963), p. 317; Amos J. Peaslee, "Political Constitution of the United States of Mexico," *Constitutions of Nations* (The Hague: Martinus Nijhoff, 1956), 2:684, 686, 687.

17. Peaslee, "Political Constitution," pp. 689–90.

1. Mexico's National Role Conceptions, 1821–1976

1. *Washington Post*, December 28, 1981, p. A-21.

2. Ibid.

3. In 1821 Mexico's land mass was 1,710 million square miles compared to 1,788 million in the United States; its population in 1810 was 6.1 million inhabitants compared to U.S. 7.2 million (Karl M. Schmitt, *Mexico and the United States, 1921–1973: Conflict and Coexistence* [New York: Wiley, 1974], pp. 44, 45, 48).

4. Francisco Cuevas Cancino, "The Foreign Policy of Mexico," in *Foreign Policies in a World of Change*, ed. Joseph E. Black and Kenneth W. Thompson (New York: Harper & Row, 1963), p. 644.

5. Ibid.

6. *Archivo Histórico-Diplomático*, 1st ser., vol. 33, p. 60, cited in ibid., p. 644.

7. William R. Manning, *Early Diplomatic Relations between the United States and Mexico* (New York: Greenwood Press, 1968), p. 164.

8. *Archivo Histórico-Diplomático*, 1st ser., vol. 36; 2d ser., vols. 4, 7.

9. Indeed, the Spanish, who were strongly affected by Islamic influence, had reinforced authoritarian and hierarchical tendencies present in indigenous cultures found in Mexico.

10. Michael C. Meyer and William L. Sherman, *The Course of Mexican History* (New York: Oxford University Press, 1983), p. 305.

11. Ibid., p. 319.

12. Frank Brandenburg, *The Making of Modern Mexico* (Englewood Cliffs, N.J., Prentice-Hall, 1964), p. 35.

13. Cuevas Cancino, "The Foreign Policy of Mexico," p. 645.

14. Eugene C. Barker, *Mexico and Texas, 1821–1835* (New York: Russell and Russell, 1965), p. 18.

15. Daniel Levy and Gabriel Székely, *Mexico: Paradoxes of Stability and Change* (Boulder, Colo.: Westview Press, 1983), p. 172; and Thomas A. Bailey, *A Diplomatic History of the American People*, 7th ed. (New York: Appleton-Century-Crofts, 1964), p. 247.

16. Bailey, *A Diplomatic History*, p. 248.

17. Levy and Székely, *Mexico*, p. 172.

18. Meyer and Sherman, *The Course of Mexican History*, p. 352.

19. Jorge Castañeda, "Revolution and Foreign Policy: Mexico's Experience," *Political Science Quarterly* 78, no. 3 (September 1963): 392.

20. Michael O'Leary, "Linkage between Domestic and International Politics in Underdeveloped Nations," in *Linkage Politics*, ed. James N. Rosenau (New York: Free Press, 1969), p. 337.

21. Quoted in Brandenburg, *The Making of Modern Mexico*, p. 321.

22. Quoted in ibid.

23. Donald D. Brand, *Mexico: Land of Sunshine and Shadow* (Princeton, N.J.: Van Nostrand, 1966), p. 147.

24. Mexico has recognized foreigners' rights to diplomatic appeal upon exhausting domestic

remedies when the denial of justice is involved; see Brandenburg, *The Making of Modern Mexico*, p. 322. See also Helen Delpar, ed., *Encyclopedia of Latin America* (New York: McGraw-Hill, 1974), pp. 103, 104, 200, 201.

25. Brandenburg, *The Making of Modern Mexico*, p. 323.

26. Maurice Halperin, "Mexico Shifts Her Foreign Policy," *Foreign Affairs* 10 (June 1941): 209–10.

27. Schmitt, *Mexico and the United States*, pp. 190–91.

28. From "Views of the Foreign Ministry of Mexico on the Dumbarton Oaks Proposals" as quoted in Castañeda, "Revolution and Foreign Policy," p. 405.

29. *United Nations Bulletin* 2 (May 13, 1947): 496.

30. Howard F. Cline, *The United States and Mexico*, p. 403.

31. Alan Riding, *Distant Neighbors: A Portrait of the Mexicans* (New York: Alfred A. Knopf, 1985), p. 342.

32. *Keesing's Contemporary Archives*, February 3–10, 1968, p. 22505; G. Pope Atkins, *Latin America in the International Political System* (New York: Free Press, 1977), pp. 366–68.

33. Olga Pellicer de Brody, *Mexico y la revolución cubana* (Mexico City: El Colegio de México, 1972).

34. Yoram Shapira, "Mexico's Foreign Policy under Echeverría: A Retrospect," *Inter-American Economic Affairs* 31 (Spring 1978): 34.

35. Guy Poitras, "Mexico's 'New' Foreign Policy," *Inter-American Economic Affairs* 28 (Winter 1974): 59–77.

36. Roberto G. Newell and Luis Rubio F., *Mexico's Dilemma: The Political Origins of Economic Crisis* (Boulder, Colo.: Westview Press, 1984), p. 110.

37. James N. Goodsell, "Mexico: Why the Students Rioted," *Current History*, January 1969, pp. 32–33.

38. Newell and Rubio, *Mexico's Dilemma*, p. 125.

39. Ibid., pp. 126, 199.

40. *El Porvenir*, September 19, 1973, p. 1; quoted in ibid., p. 200.

41. Shapira, "Mexico's Foreign Policy under Echeverría," p. 34.

42. Ibid., pp. 42–43.

43. Ibid., p. 30.

44. Ibid., p. 34.

45. On diminished U.S. influence, see Bruce Michael Bagley, "Mexican Foreign Policy in the 1980's: A New Regional Power," *Current History* 80 (November 1981): 353–56; and Edwin A. Deagle, Jr., "United States National Security Policy and Mexico," in *U.S.–Mexican Relations: Economic and Social Aspects*, ed. Clark W. Reynolds and Carlos Tello (Stanford, Calif.: Stanford University Press, 1983), pp. 200–01.

46. Bruce Michael Bagley, "Regional Powers in the Caribbean Basin: Mexico, Venezuela, and Colombia," Occasional Paper No. 2, School of Advanced International Studies, Johns Hopkins University, Washington, D.C., January 1983, pp. 3–4.

47. *Washington Post*, March 18, 1975, p. 15.

48. *New York Times*, November 20, 1976, p. 8; November 23, 1976, p. 54; November 26, 1976, pp. 1–10.

2. Oil and the Economic Crises

1. See George W. Grayson, *The Politics of Mexican Oil* (Pittsburgh, Pa.: University of Pittsburgh Press, 1980); Richard B. Mancke, *Mexican Oil and Natural Gas: Political, Strategic,*

and Economic Implications (New York: Praeger, 1979); and Edward J. Williams, *The Rebirth of the Mexican Petroleum Industry* (Lexington, Mass.: Lexington Books, 1979).

2. The source of most data on oil activities is Petróleos Mexicanos, *Anuario estadístico 1983* (Mexico City: Petróleos Mexicanos).

3. For a discussion of the inflated reserve figures released by the Mexican government, see Grayson, *The Politics of Mexican Oil*, p. 70ff.

4. *Comercio exterior de méxico* 26, no. 3 (March 1980): 81–82; *Forbes*, October 29, 1979, pp. 42–44.

5. José López Portillo, *Cuarto informe de gobierno* (Mexico City: Presidencia de la República, 1980), p. 63.

6. Jean-Jacques Servan-Schreiber, *Le défi mondial* (Paris: Fayard, 1980), p. 72.

7. This section draws on material in George W. Grayson, *The United States and Mexico: Patterns of Influence* (New York: Praeger, 1984), ch. 4.

8. Bagley, "Regional Powers in the Caribbean Basin," p. 1.

9. *Fortune*, July 16, 1979, p. 138.

10. *Newsweek*, October 1, 1979, p. 26.

11. *Fortune*, July 23, 1982, p. 151.

12. *Latin America Economic Report*, November 4, 1977, p. 197.

13. José López Portillo, *Second State of the Nation Report* (Mexico City: Presidency of the Republic, 1978), p. 147.

14. *Facts on File*, October 5, 1979, p. 753.

15. Foreign Broadcast Information Service, *Daily Report (Latin America)*, March 20, 1980, p. M-2.

16. *New York Times*, November 14, 1979, pp. A-1, A-24; Herbert E. Meyer, "Why a North American Common Market Won't Work—Yet," *Fortune*, September 10, 1979, pp. 118–24; Wallace C. Koehler, Jr., and Aaron L. Segal, "Prospects for North American Energy Cooperation," *USA Today*, May 1980, pp. 40–43; George W. Grayson, "The Maple Leaf, the Cactus, and the Eagle: Energy Trilateralism," *Inter-American Economic Affairs* 34 (Spring 1981): 49–75.

17. Mario Ojeda, "The Future of Mexico-U.S. Relations," in *U.S.-Mexico Relations: Economic and Social Aspects*, ed. Clark W. Reynolds and Carlos Tello (Stanford, Calif.: Stanford University Press, 1983), p. 323.

18. *Tiempo*, February 21, 1977, p. 8.

19. *Comercio exterior de méxico* 29, no. 2 (February 1979): 160.

20. *New York Times*, May 18, 1979, p. A-3.

21. Ibid.; *Washington Post*, May 18, 1979, p. A-31.

22. *New York Times*, May 18, 1979, p. A-3.

23. *Regional Reports: Mexico and Central America*, August 15, 1980, p. 4.

24. Jorge G. Castañeda, "Don't Corner Mexico!" *Foreign Policy* 60 (Fall 1985): 82.

25. Bagley, "Regional Powers in the Caribbean Basin," p. 24.

26. Ibid., p. 13.

27. Castañeda, "Don't Corner Mexico!" pp. 80–81.

28. Bagley, "Regional Powers in the Caribbean Basin," p. 12.

29. Castañeda, "Don't Corner Mexico!" p. 80.

30. José López Portillo, *Tercer informe de gobierno* (Mexico City: Presidencia de la República, 1979), p. 82.

31. Castañeda, "Don't Corner Mexico!" pp. 80–81.

32. López Portillo, *Tercer informe*, p. 82.

33. *Excelsior*, January 25, 1980, pp. 1-A, 10-A; López Portillo, *México en el ambito internacional* 4 (Mexico City: Secretaria de Programación y Presupuesto, 1981), pp. 9–13.

34. The organization is discussed in "COPPPAL: An Appreciation of the Regional Political Situation," *Comercio exterior de méxico* 28, no. 5 (May 1982): 173–79.

35. Constantine Christopher Menges, "Concurrent Mexican Foreign Policy, Revolution in Central American and the United States," presented at the Hudson Institute conference, Washington, D.C., June 1980, p. 43.

36. *New York Times,* August 20, 1980, p. 8-A.

37. *Washington Post,* August 18, 1980, p. 19; Castañeda, "Don't Corner Mexico!" p. 81.

38. *Facts on File,* March 6, 1981, p. 135.

39. The text of the document appears in *Keesing's Contemporary Archives,* November 6, 1981, p. 31174.

40. Undated memorandum from Cheryl Eschbach, Ph.D. candidate in the Department of Government, Harvard University.

41. Presidencia de la República, *Plan mundial de energía: proposición ante las naciones unidas* (Mexico City: Coordinación General de Comunicación Social, 1979).

42. López Portillo, *Tercer informe,* p. 87.

43. Because of illness, Kreisky was replaced as co-chairman of the meeting by Canadian Prime Minister Trudeau; see Partido Revolucionario Institucional, *Diálogo norte sur* (Mexico City: Tallares Gráficos de la Nación, 1981).

44. López Portillo, *Sexto informe de gobierno* (Mexico City: Presidencia de la República, 1982), p. 13.

45. *New York Times,* April 24, 1980, p. 2.

46. Roberto G. Newell and Luis Rubio F., *Mexico's Dilemma: The Political Origins of Economic Crisis* (Boulder, Colo.: Westview, 1984), 207–08.

47. Alan Riding, "The Mixed Blessings of Mexico's Oil," *New York Times Magazine,* January 11, 1981, p. 25.

48. Wayne A. Cornelius, "The Political Economy of Mexico under de la Madrid: Austerity, Routinized Crisis, and Nascent Recovery," *Mexican Studies/Estudios Mexicanos* 1, no. 1 (Winter 1985): 91.

49. Joseph Kraft, *The Mexican Rescue* (New York: Group of Thirty, 1984), p. 37.

50. *Wall Street Journal,* September 15, 1982, p. 36.

51. The contract covering Pemex's sale to the reserve, which I obtained from the Department of Energy under the Freedom of Information Act, indicates that 40,100,000 barrels of crude were delivered for the exceedingly attractive price of slightly more than $25 per barrel; see U.S. Department of Energy Contract No. DE-AC01-82EP31403, signed August 24, 1982.

52. *Excelsior,* September 2, 1982, p. 1-A.

53. *Latin America Weekly Report,* November 19, 1982, p. 6.

54. *Washington Post,* September 26, 1982, p. A-20.

55. *Excelsior,* August 6, 1982, pp. 1-A, 9-A, 11-A; August 18, 1982, pp. 1-A, 18-A, 30-A; September 2, 1982, pp. 1-A, 18-A, 20-A, 28-A, 30-A.

56. *Excelsior,* September 2, 1982, p. 28-A.

57. Castañeda, "Don't Corner Mexico!" p. 84.

58. López Portillo, *Sexto informe,* p. 18.

59. In an unofficial statement in late October 1986, Jesús Salazar Toledano, a leader of the PRI in Mexico City, "uncovered" the names of four men from whom he believed the presidential candidate would be selected: Energy Secretary Alfredo del Mazo; Interior Secretary Manuel Bartlett Díaz; Planning and Budget Secretary Carlos Salinas de Gotari; and Education Secretary Miguel González Avelar. See *Latin American Weekly Report,* November 6, 1986, p. 10. Subsequently, the names of Attorney-General Sergio García Ramírez and Mexico City's Mayor Ramón Aguirre Velázquez were added to the list of the party's official precandidates.

60. The text of the speech appears in Foreign Broadcast Information Service, *Daily Report (Latin America)*, December 2, 1982, pp. M-2, M-10.

61. *New York Times*, October 12, 1982, p. D-17; *Washington Post*, October 8, 1982, p. D-8; Bagley, "Regional Powers in the Caribbean Basin," pp. 18–19; World Bank, *World Development Report 1986* (Washington, D.C.: Oxford University Press, 1986).

62. Quoted in Kraft, *The Mexican Rescue*, p. 1; *Economist*, January 29, 1983, p. 60.

63. Cornelius, "The Political Economy of Mexico," p. 109.

64. *Wall Street Journal*, December 4, 1985, p. 35.

65. For articles on the Contadora peace effort, see Bruce Michael Bagley, "Contradora: The Failure of Diplomacy," *Journal of InterAmerican Studies and World Affairs* 28 (Fall 1986): 1–32; Susan Kaufmann Purcell, "Demystifying Contadora," *Foreign Affairs* 64 (Fall 1985): 74–95.

66. Bagley, "Contadora," p. 2.

67. Castañeda, "Don't Corner Mexico!" p. 86.

68. *Los Angeles Times*, October 31, 1985, sec. 2, p. 7.

69. *Financial Times*, March 3, 1986, p. 1.

70. *Latin American Weekly Report*, October 16, 1986, pp. 1, 7.

71. Jack Sweeney, "Finance to Flow from Accord . . . Against Promises," *Business Mexico* 4 (March 1987), p. 26.

72. *Facts on File*, May 18, 1984, pp. 346–47.

73. Anatole Kaletsky, *Financial Times*, July 25, 1986, p. 16.

74. See Norman A. Bailey and Richard Cohen, *The Mexican Time Bomb* (New York: Priority Press, 1987).

3. Mexico's Oil Industry Under de la Madrid: A New Pemex?

1. Corruption is discussed in Angelina Alonso Palacios and Carlos Roberto López, "El sindicato de trabajadores petroleros y sus relaciones con pemex y el estado, 1970–1982," Cuadernos sobre prospectiva energética, El Colegio de México, Mexico City, n.d., mimeo. See also, George W. Grayson, *The Politics of Mexican Oil*, ch. 4, and "An Overdose of Corruption: The Domestic Politics of Mexican Oil," *Caribbean Review* 13, no. 3 (Summer 1984): 22–24, 46–49.

2. Quoted in José Camacho Morales, *El nuevo pemex* (Mexico City: Petróleos Mexicanos, 1983), p. 133.

3. Petróleos Mexicanos, *General Director's Report: Petróleos Mexicanos* (Mexico City: Mexican Petroleum Institute, 1983), p. 9.

4. Ibid., p. 21.

5. Ibid., p. 10.

6. Poder Ejecutivo Federal, *Programa nacional de energéticos, 1984–1988* (Mexico City: SEMIP, 1984); New York/Washington Representative Offices of Petróleos Mexicanos, *Pemex: Information Bulletin* 12 (September 1984): 1–3; ibid., no. 16 (January 1985): 7; *Platt's Oilgram News*, August 15, 1984, p. 2.

7. The two subdirectors recruited from within the monopoly were Fernando Manzanilla Sevilla (Projects and Construction) and Miguel Angel Zenteno Basurto (Primary Production), who—following his mid-1984 death in an airplane crash—was succeeded by Abundio Juárez Méndez. The former Somex executives were Alberto Bremauntz Monge (Industrial Transformation), Flavio Pérezgasga Tovar (Planning and Coordination), Donaciano Támez Fuentes (Commercial), and Roberto Morales Martínez, who was succeeded by Rodolfo Echeverría Ruiz (Technical Administrative). Joaquín Muñoz Izquierdo, the finance subdirector, had previously served as director of finance and administration for Teléfonos de México, S.A., the state monopoly. Upon assuming the helm of Pemex in early 1987, Francisco Rojas replaced four subdirectors. His appointees were

Juan Enrique Vázquez Domínguez (Industrial Transformation), Roberto Morales Martínez (Planning and Coordination), Raul Robles Segura (Commercial), and Cuauhtémoc Santa Ana Seuthe (Technical Administrative).

8. *Business Week,* February 28, 1983, p. 61.

9. Interview with Ing. Flavio Pérezgasga Tovar, subdirector for planning and coordination, Petróleos Mexicanos, August 15, 1985, Mexico City.

10. *Platt's Oilgram Price Report,* May 18, 1987, p. 3-A; *Pemex: Information Bulletin* 43 (April 1987): 8.

11. Petróleos Mexicanos, *A Report by Petróleos Mexicanos to Mr. Miguel de la Madrid, the President of Mexico, about the Industry's Progress and Prospects during the First Two Years of His Administration* (Mexico City: Pemex, n.d.), p. 10.

12. Petróleos Mexicanos, *Report from the General Director of Petróleos Mexicanos: Mario Ramón Beteta* (Mexico City: Mexican Petroleum Institute, 1985), pp. 19–20; *A Report by Petróleos Mexicanos to Mr. Miguel de la Madrid,* p. 12; Petróleos Mexicanos, *Third Evaluation Meeting* (Mexico City: Mexican Petroleum Institute, 1987), p. 76.

13. Banamex, *Review of the Economic Situation in Mexico* 42, no. 726 (May 1986): 179.

14. Petróleos Mexicanos, *Report Presented by the Director General of Petróleos Mexicanos* (Mexico City: Mexican Petroleum Institute, 1986), p. 8.

15. *Pemex: Information Bulletin* 27 (December 1985): 8.

16. Petróleos Mexicanos, *Third Evaluation Meeting,* p. 30.

17. Petróleos Mexicanos: *Report from the General Director* (1985), pp. 3–4; *Report Presented by the Director General* (1986), p. 7; *Memoria de labores 1986* (Mexico City: Mexican Petroleum Institute, 1987), p. 3.

18. *Pemex: Information Bulletin* 19 (February 1986): 3.

19. Grayson, *Politics of Mexican Oil,* pp. 71–72; see also an excellent paper on how Pemex has inflated its reserve figures by one-third or more by Walter Friedeberg M., former manager for primary production, Pemex, and now a professor in UNAM's School of Engineering: "Las reservas de hidrocarburos y la capacidad productiva como base para la planeación de la explotación petrolera," April 1986, mimeo.

20. *A Report by Petróleos Mexicanos to Mr. Miguel de la Madrid,* p. 56.

21. *Pemex: Information Bulletin* 16 (January 1985): p. 6; and ibid., p. 11.

22. Ibid., pp. 6–7; Petróleos Mexicanos: *Report Presented by the Director General* (1986), p. 14; *Third Evaluation Meeting,* p. 70.

23. See Grayson, *Politics of Mexican Oil,* ch. 9.

24. Petróleos Mexicanos, *Report from the General Director* (1985), p. 24.

25. *Washington Post,* November 26, 1984, p. A-18.

26. *New York Times,* December 6, 1984, p. A-12.

27. *Proceso,* May 11, 1987, pp. 32–33.

28. Petróleos Mexicanos, *General Director's Report* (1983), p. 19.

29. Grayson, "An Overdose of Corruption," p. 48.

30. *Proceso,* September 3, 1984, p. 13.

31. *Diario Oficial,* January 30, 1984, pp. 8–10.

32. *Latin America Regional Reports: Mexico & Central America Report,* May 4, 1984, p. 4; and *Platt's Oilgram News,* February 9, 1984, p. 2. The union also owns several drilling companies, including Aguila and Perforadora Veracruz.

33. *New York Times,* February 6, 1984, p. A-7.

34. Ibid.

35. *Wall Street Journal,* January 20, 1984, pp. 1, 12.

36. *Platt's Oilgram News,* February 9, 1984, p. 2.

37. *Wall Street Journal*, January 20, 1984, p. 1.

38. Quoted in *Proceso*, November 26, 1984, p. 13.

39. Héctor Aguilar Camín, *Morir en el golfo* (Mexico City: Ediciones Oceano, 1987).

40. Jaime Aguilar Briseño, *"La Quina"* (Mexico City: Editorial Color, 1980); Rafael Ramírez Heredia, *La otra cara del petróleo* (Mexico City: Editorial Diana, 1979).

41. *Proceso*, July 23, 1984, p. 7.

42. Another extraordinarily flattering article, adorned by thirteen color photographs, was entitled "Una labor social sin precedentes del trabajador petrolero"; see *Siempre!* June 20, 1984, pp. 43–46.

43. *Proceso*, November 26, 1984, p. 6.

44. Ibid., October 22, 1984, p. 9.

45. *Proceso* published the names of 240 employees included in *nóminas*, each signed by the appropriate subdirector and authorized by the technical administrative subdirector, of the Primary Production, Commercial, and Planning and Coordination Subdirectorates. Although names were added later, these lists dated to January 1983, one month after Beteta became Pemex's head (ibid.).

46. Ibid.

47. Francisco Arellano-Belloc T., manager of social services, received 198,005.81 pesos in October 1983 for the treatment of his wife, Patricia Jiménez de Arellano-Belloc, in the American British Cowdray Hospital; see ibid.

48. According to a memorandum by Georgel Rosales Alvarado, internal manager of technical administration services, the monopoly paid 975,977 pesos to rent an automobile (January 26 and June 15, 1983) for Gilberto Escobedo Villalon. This vehicle was not returned when the former commercial subdirector left Pemex for the Banco Mexicano Somex; see ibid.

49. Fifteen Pemex employees (six permanent and nine contract or transitory) were assigned to the PRI; one (María González Ayon) to the PPS; see ibid., September 3, 1984, pp. 13, 14, 16.

50. Involved in this relocation, the cost of which reportedly exceeded 100 million pesos, was the Executive Coordination Office for Supply and the Purchase Agency whose office space is located at the corner of Homero and Taine Streets; see ibid., September 24, 1984, pp. 21–22.

51. Payments totaling 28,268,861 pesos were authorized to the firm Realizaciones Industriales, S.A., for remodeling offices on the twelfth floor of the Torre Pemex when the real cost of the work was 16,119,086, according to an internal report of the *gerencia* of Technical Administrative Services (77111-258/83); see ibid., September 3, 1984, p. 12.

52. In mid-1984, Industrias Papanoa, S.A., was awarded a contract for 449.4 million pesos to construct a hotel for pilots in Ciudad Carmen, near Pemex's largest offshore operations. Fernando Manzanilla Sevilla, Subdirector for Projects and Construction, cited Article 30 of the Public Works Law to justify the lack of competition because Papanoa held the necessary patents to build the prefabricated structure. Humberto Mota, owner of the firm Construcciones California, claimed "favoritism" in asserting that his company also possessed the capability to do the work. See ibid., pp. 14–15.

53. By virtue of one document, Pemex forgave the SRTPRM a 78-million-peso debt and awarded it an additional 132 million pesos; see ibid., pp. 13, 16, 17.

54. Both quotations appear in the *Wall Street Journal*, October 9, 1985, p. 26.

55. Ibid., January 20, 1984, p. 1.

56. Quoted in *Proceso*, July 23, 1984, p. 9–10.

57. Ibid., p. 6.

58. *El Universal*, June 2, 1986; and *Proceso*, June 9, 1986, pp. 6–11.

59. Major accidents in 1986 included: a fire in a gas-processing plant in Cactus (January 11), a fire in the Santa Cruz refinery (January), a fire in the Luna II well in Tabasco state (September 20), and a fire in the Abkatún 91 well (October 23). No loss of life occurred in the four fires, all of which

were extinguished after substantial property damage; see *Platt's Oilgram News,* January 14, 1986, p. 4, January 28, 1986, p. 4, October 1, 1986, p. 3, October 28, 1986, p. 1, November 11, 1986, p. 6.

60. Petróleos Mexicanos, *Third Evaluation Meeting,* p. 57.

61. *Pemex: Information Bulletin* 44 (May 1987): 8; Francisco Rojas, *Hermanemos la idea a la acción* (Mexico City: Mexican Petroleum Institute, 1987), p. 6.

62. *Wall Street Journal,* October 9, 1985, p. 26.

63. This estimate came from Mexico's Energy Ministry; see *Financial Times,* July 12, 1985, p. 1.

64. *Wall Street Journal,* July 12, 1985.

65. *Business Week,* February 28, 1983, p. 58.

4. United States–Mexican Relations

1. *Washington Post,* May 14, 1986, p. A-3.

2. Ibid.

3. *Facts on File,* May 30, 1986, p. 399.

4. On an ABC news program, Meese called the first Helms hearing "reckless" and said the charges by "one of the unfortunate people in the Customs Service" had given the impression that the whole Mexican government was involved in drug trafficking. The attorney general also denied the accusations that von Rabb had made against Sonora's governor, Rodolfo Félix Valdés (*This Week with David Brinkley,* May 15, 1986).

5. Peter H. Smith, "Uneasy Neighbors: Mexico and the United States," *Current History* 86 (March 1987): 97.

6. For a discussion of technocrat de la Madrid's establishing rapport with the people, see George W. Grayson, "Middle-Class Agitation in Quaking Mexico," *Washington Post,* December 8, 1985, pp. B-1, B-4.

7. *Excelsior,* June 3, 1986, p. 9-A.

8. *Pemex: Information Bulletin* 15 (December 1984): 2.

9. Ibid. 22 (July 1985): 1–3.

10. As an example of the expertise transmitted through the consultative mechanism, see the names (and topics) of the several DOE specialists who addressed the April 24–25, 1986, session of the BECG; ibid. 32 (June 1986): 2–3.

11. These negotiations are discussed in George W. Grayson, "The U.S.–Mexican Gas Deal and What We Can Learn from It," *Orbis* 78, no. 454 (February 1980): 53–56, 83; see also Richard R. Fagen and Henry R. Nau, "Mexican Gas: The Northern Connection," presented at the Conference on the United States, U.S. Foreign Policy, and Latin American and Caribbean Regimes, Washington, D.C., March 27–31, 1978.

12. In addition to Tennessee Gas that agreed to take 37.5 percent of the gas, the consortium included Texas Eastern Transmission Corporation (27.5 percent), El Paso Natural Gas Company (15 percent), Transcontinental Gas Pipeline Corporation (10 percent), Southern Natural Gas Company (6.7 percent), and Continental Resources, Inc. (3.3 percent).

13. $\text{Gas Price} = \dfrac{A \text{ (average of 5 crudes)}}{27.44 \text{ (crude price at time of accord)}} \times 3.625 \text{ (initial contract price)}.$

14. New gas came from wells that had begun producing after April 30, 1977.

15. *Wall Street Journal,* May 27, 1983, p. 26.

16. Ibid.

17. Ibid.

18. *Facts on File,* July 22, 1983, p. 551.

19. *Pemex: Information Bulletin* 15 (December 1984): 4.

20. Text of a press release drafted by Border Gas for its use on Wednesday, Oct. 24, 1984, Houston, Texas.

21. Dianne Klein, "Repainting the Picture: Mexico Turns to PR to Improve Image in U.S.," *Houston Chronicle,* April 20, 1986, p. 30.

22. David Ronfeldt and Caesar Sereseres, "Immigration Issues Affecting U.S.–Mexican Relations," presented at the Brookings Institution–El Colegio de México Symposium on Structural Factors Contributing to Current Patterns of Migration in Mexico and the Caribbean Basin, Washington, D.C., June 28–30, 1978, p. 9.

23. Interview with Ambassador Robert E. Fritts, diplomat-in-residence, College of William and Mary, Williamsburg, Virginia, September 8, 1986.

24. *Wall Street Journal,* July 29, 1985, p. 1.

25. Ibid., p. 8.

26. As it turned out, Deaver began discussing a "possible contract" with Canada on May 16, 1985—six days after he left government service. See *Wall Street Journal,* May 12, 1986, p. 12.

27. *Washington Post,* May 12, 1986, pp. B-1, B-4.

28. *Wall Street Journal,* April 11, 1985, p. 34.

29. Joseph Blatchford, quoted in the *Wall Street Journal,* April 11, 1985, p. 34.

30. Alfred Gutiérrez Kirchner, "Presencia, funciones y actividades de petróleos mexicanos en los estados unidos de america," 1983, mimeo., p. 11.

31. Ibid. p. 15.

32. The mansion, located on 71st Street between Fifth and Madison Avenues, belonged to a textile tycoon until the 1920s when the Diocese of New York opened a hospital for blind children. An additional $2 million would have been required to renovate its four floors and penthouse. Crédit Lyonnais financed the Pemex purchase, giving rise to a $50,000 per month interest payment during a period of economic retrenchment. By advertising in the *New York Times,* Gutiérrez Kirchner managed to sell the building for $7 million in early 1983. He also reduced the number of employees in the New York office from forty-six to ten.

33. *Pemex: Information Bulletin* 9 (June 1984): 5.

34. Ibid. 11 (August 1984): 7.

35. Ms. Sue Sarnio, research assistant to George W. Grayson, conducted the telephone interviews.

36. See, for example, the "U.S. Council of the Mexico–U.S. Business Committee, 1984–85 Report," prepared by the Council, New York, n.d., mimeo.

37. On the accord, see "Communiqué," Mexico–U.S. Business Council, 41st Plenary Meeting, Quintana Roo, Mexico, November 8, 1986 (draft communiqué dated January 21, 1981).

38. Klein, "Repainting the Picture," p. 30.

39. Ibid.

40. Published by Coward-McCann, Inc., New York, 1984.

41. In return for $250,000 annually, Michael K. Deaver and Associates were to provide research, public opinion surveys, the development of a communications strategy, lobbying, and an analysis of political and commercial developments in the United States relevant to Mexican interests; see *Proceso,* May 5, 1986, pp. 22–24.

42. Klein, "Repainting the Picture," p. 30.

43. Interview with Mark E. Moran, general counsel, The Hannaford Company, Washington, D.C., May 22, 1986.

44. Klein, "Repainting the Picture," p. 30.

45. Hannaford Company news releases: "Camarena Slaying Suspects Seized as Mexico's Drug

Crackdown Continues," issued February 6, 1986; "Confession Implicates Camarena Slaying Suspects," news release issued March 27, 1986; "Mexico Police Net Record $80 Million Drug Haul," news release issued April 9, 1986.

46. Letter from Jared S. Cameron dated July 23, 1986.

47. Secretariat of Information and Propaganda, Institutional Revolutionary Party, *Who is Miguel de la Madrid*[?] (Mexico City: PRI, 1982).

48. Interview with Moran, July 24, 1986, Washington, D.C., November 11, 1987, Mexico City.

5. Mexico and the Organization of Petroleum Exporting Countries

1. *Platt's Oilgram News*, May 6, 1976, p. 1; Edward J. Williams, "Mexico, Oil, and OPEC," *Latin American Digest* 11–12 (Fall–Winter 1977–1978): 6; and *Middle East Economic Survey*, May 10, 1976, p. 2.

2. *Platt's Oilgram News*, September 24, 1975, p. 2.

3. *Middle East Economic Survey*, July 25, 1975, p. 6.

4. *Excelsior*, September 10, 1977, p. 4-A.

5. *Platt's Oilgram News*, March 27, 1978, p. 4.

6. *Oil Daily*, November 3, 1978, p. 8.

7. *Daily Report (Latin America)*, June 20, 1978, p. M-1.

8. *Excelsior*, November 3, 1977, p. 5-A; August 30, 1979, p. 17-A.

9. *Washington Post*, July 1, 1979, p. A-30.

10. *Daily Report (Latin America)*, March 30, 1979, p. M-1.

11. *Middle East Economic Survey*, May 14, 1976, p. 7.

12. *Proceso*, February 12, 1979, p. 9.

13. Presidencia de la República, *Plan mundial de energía: proposición ante las naciones unidas* (Mexico City: Coordinación General de Comunicación Social, n.d.), p. 41.

14. Ibid., pp. 43–44.

15. Ibid., pp. 34, 35, 39, 45.

16. Ibid., p. 47.

17. Ibid., pp. 48–49.

18. Ibid., p. 51.

19. Jimmy Carter, *Keeping Faith: Memoirs of a President* (New York: Random House, 1982), pp. 452–56, 468.

20. *Proceso*, October 15, 1979, p. 42.

21. Ibid.

22. *Platt's Oilgram News*, October 12, 1979, p. 3.

23. *Latin American Weekly Report*, November 12, 1982, p. 1.

24. *Petroleum Intelligence Weekly*, June 28, 1982, p. 1.

25. Ibid.

26. *New York Times*, January 28, 1983, sec. 4, p. 1.

27. Partido Revolucionario Institutional, Instituto de Estudios Políticos, Económicas y Sociales, Comisión de Energéticos, "Energía y sector externo" (draft for discussion), September 22, 1982, p. 22; and *Latin American Weekly Report*, November 12, 1982, pp. 1–2.

28. PRI, "Energía y sector externo," p. 23.

29. Heavy crudes were independent of OPEC's price structure but Mexico reduced Maya to $23 per barrel, aligning it with the Venezuelan price for heavy varieties. In August 1983, both countries raised their heavy crude prices to $24; later Pemex increased its price for Maya to $25.

30. *Petroleum Intelligence Weekly*, May 30, 1983, p. 9.

31. *Petroleum Economist* 50, no. 5 (May 1983): 265.

32. *Wall Street Journal,* March 8, 1984, p. 35.

33. *OPEC Bulletin* 15, no. 3 (April 1984): 15.

34. *Business Week,* August 15, 1983, p. 29.

35. *Wall Street Journal,* January 10, 1985, p. 22.

36. *Daily Report (Latin America),* December 7, 1983, p. M-1.

37. *Platt's Oilgram News,* June 24, 1983, p. 1.

38. *Wall Street Journal,* August 2, 1983, p. 32.

39. *Daily Report (Latin America),* November 29, 1983, p. A-1.

40. *Wall Street Journal,* March 8, 1984, p. 35.

41. *OPEC Bulletin* 15, no. 3 (April 1984): 48–54.

42. *Platt's Oilgram News,* July 30, 1984, p. 3.

43. *Pemex: Information Bulletin* 14 (November 1984): 2.

44. *Latin America Regional Report: Mexico and Central America,* February 15, 1985, p. 2.

45. *Platt's Oilgram News,* January 15, 1984, p. 4.

46. *Latin America Regional Report: Mexico and Central America,* February 15, 1985, p. 2;
Pemex: Information Bulletin 16 (January 1985): 4.

47. *Wall Street Journal,* July 9, 1984, p. 29.

48. Ibid., July 10, 1985, p. 2.

49. *Business Week,* February 25, 1985, p. 46.

50. *Middle East Economic Survey,* December 31, 1984, pp. 3–4.

51. *Economist,* January 19, 1985, p. 65.

52. *Platt's Oilgram News,* January 22, 1985, p. 1.

53. *Wall Street Journal,* January 22, 1985, p. 6.

54. *Latin America Regional Report: Mexico and Central America,* February 15, 1985, p. 2.

55. *Oil and Gas Journal,* February 11, 1985 (newsletter), p. 1.

56. *Pemex: Information Bulletin* 17 (February 1985): 2.

57. *Latin America Regional Report: Mexico and Central America,* February 15, 1985, p. 2.

58. *Economist,* January 19, 1985, p. 65; and *Petroleum Intelligence Weekly,* January 14, 1985,
p. 7.

59. *Economist,* January 19, 1985, p. 66.

60. Gabriel Székely, "México y el petróleo: 1981–1985," *La Jornada,* September 1, 1985,
p. 5.

61. *Wall Street Journal,* January 31, 1985, p. 33.

62. Ibid., February 3, 1986, p. 3.

63. *Financial Times,* March 1986.

64. *Pemex: Information Bulletin* 40 (January 1987): 1.

6. Mexico and the San José Accord

1. Edward F. Wonder and J. Mark Elliott, "Caribbean Energy Issues and U.S. Policy," in
Western Interests and U.S. Options in the Caribbean Basin, ed. James R. Greene and Brent
Scowcroft (Boston: Oelgeschlager, Gunn, & Hain, 1984), pp. 269–304.

2. Robert D. Bond, "Venezuelan Policy in the Caribbean Basin," in *Central America; International Dimensions of the Crisis,* ed. Richard E. Feinberg (New York: Holmes & Meier, 1982), p.
192.

3. Ibid.

4. The plan is analyzed in George W. Grayson, "Venezuela and the Puerto Ordaz Agreement,"
Inter-American Economic Affairs 38, no. 3 (Winter 1984): 49–73.

5. Ministerio de Información y Turismo, *La cooperación internacional de venezuela: solidaridad en acción* (Caracas: Imprenta Nacional, 1982), p. 15.

6. Petróleos Mexicanos, *Anuario estadístico 1980* (Mexico City: Instituto Mexicano de Petróleo, 1980), p. 29, 37.

7. *Daily Report (Latin America),* April 5, 1979, p. M-1.

8. Presidencia de la República, *Plan mundial de energía.*

9. López Portillo, *Quinta informe de gobierno,* p. 65.

10. *Excelsior,* January 25, 1980, pp. A-1, A-10.

11. *Proceso,* February 11, 1980, p. 26; *Petroleum Intelligence Weekly,* February 18, 1980, p. 11.

12. For a description of the essential elements in the San José Accord, see Ministry of Energy and Mines, *La cooperación energetica de venezuela / The Energy Cooperation of Venezuela* (Caracas: Ediciones del Ministerio de Energía y Minas, 1981), pp. 67–73.

13. Venezuela's "cocktail" is a blend of approximately 45 percent crude plus refined products such as light oil, diesel oil, gas oil, kerosene, which is designed to yield a refined product slate precisely tailored to the internal demand of a specific country when processed through its existing refinery(ies) which may not be sophisticated enough to generate sufficient light product from crude alone. Only Venezuela exports "recon," often blended to balance Mexican crude to achieve desired refinery throughput. Panama, which has a more sophisticated refinery, is the only facility client to which PDVSA supplies natural crude.

14. Ministry of Energy and Mines, *La cooperación energética de venezuela,* p. 69.

15. *Petroleum Intelligence Weekly,* April 13, 1981, p. E.

16. "Caribbean Oil Deal," *South,* November 1980, p. 56.

17. The contracting entities are as follows: Barbados (Ministry of Finance), Costa Rica (Recope), Dominican Republic (Presidency of the Republic), El Salvador (Comisión Eléctrica Río Lempa), Guatemala (Ministry of Energy and Mines), Honduras (Ministry of Energy), Jamaica (Petrojam), Nicaragua (Ministry of Economy), and Panama (Ministry of Trade and Industry).

18. "Trinidad's Oil Facility Gets Underway," *Caribbean and West Indies Chronicle* 96, no. 1557 (1980): 15.

19. For a summary of disbursements under the program, see *Trinidad & Tobago's Economic Assistance to the Caricom Member Countries: a Brief Review–1979–1982* (Washington, D.C.: Embassy of Trinidad and Tobago, n.d.)

20. Interviews with Dr. Julio Gil, minister-counselor, Embassy of Venezuela, Washington, D.C., June 5, 1984; and with René Arreaza, Petróleos de Venezuela, S.A., Caracas, August 7, 1984.

21. Interview with Valentín Hernández, Venezuelan ambassador to the United States and Minister of Energy and Mines (1974–1979), Washington, D.C., May 30, 1984.

22. Minister of Energy and Mines, *La cooperación energética de venezuela,* p. 70.

23. U.S. Department of State, "The Mexican-Venezuelan Oil Facility—an Assessment," Washington, D.C.: Department of State, May 13, 1983, mimeo., p. 2.

24. *Washington Post,* December 25, 1981, p. A-2.

25. *Petroleum Intelligence Weekly,* October 27, 1986, p. 8.

26. Among the facility nations, Texaco has three refineries (Guatemala, Honduras, and Panama), Esso two (El Salvador and Nicaragua), and Mobil (Barbados), Recope (Costa Rica), Refilonsa (Dominican Republic), and Petrojam (Jamaica) one apiece.

27. *Petroleum Intelligence Weekly,* April 13, 1981, p. 3.

28. U.S. Department of State, "The Oil Facility in 1982 and Prospects for 1983," Washington, D.C., Department of State, May 13, 1983, mimeo., pp. 2–3.

29. Dan Weil, "Mexico, Venezuela Seen Likely to Renew San José Accord Although its Significance has Lessened," AP–Dow Jones Wire Service, July 17, 1986, p. 2.

30. *Petroleum Intelligence Weekly,* July 30, 1984, p. 4.

31. *Excelsior,* August 4, 1986, pp. 1-A, 14-A, 39-A.

32. *Platt's Oilgram News,* May 8, 1986.

33. U.S. Department of State, Bureau of Public Affairs, *Caribbean Basin Initiative,* Current Policy No. 370, Washington, D.C., 1982; see also U.S. Department of State, Bureau of Public Affairs, *Background on the Caribbean Basin Initiative,* Special Report No. 97, Washington, D.C., 1982.

34. *Latin America Weekly Report,* July 17, 1981, p. 11.

7. Mexico and Nicaragua

1. Thomas A. Walker, "Nicaragua: The Somoza Family Regime," in *Latin American Politics and Development,* ed. Howard J. Wiarda and Harvey F. Kline (Boston: Houghton Mifflin, 1979), pp. 324–25.

2. *Excelsior,* January 15, 1980, pp. 1-A, 10-A; *Latin America Regional Report: Mexico and Central America,* February 15, 1980, p. 4.

3. See the *International Petroleum Encyclopedia,* 1965–1976 (Tulsa, Okla.: Petroleum Publishing Co.).

4. *World Oil,* August 15, 1979, p. 94.

5. *Oil and Gas Journal,* January 21, 1980, p. 54.

6. *World Oil,* August 15, 1980, p. 94; August 15, 1983, p. 53.

7. Ibid., August 15, 1982, p. 97.

8. *New York Times,* May 21, 1985, p. A-10.

9. *International Petroleum Encyclopedia,* 1973, p. 155.

10. *Petroleum Economist* 50, no. 11 (November 1983): 437.

11. *Daily Report (Latin America),* May 8, 1984, p. M-2.

12. *Petroleum Economist* 51, no. 5 (May 1984): 190.

13. *Platt's Oilgram News,* April 4, 1984, pp. 1–2.

14. *Daily Report (Latin America),* April 18, 1984, p. M-4.

15. Ibid., May 23, 1985, p. M-4.

16. Ibid., February 4, 1983, p. M-3.

17. *New York Times,* August 13, 1983, p. 3.

18. *Daily Report (Latin America),* August 22, 1983, pp. P-19, P-20.

19. Ibid., August 29, 1983, p. P-24.

20. Ibid., February 14, 1984, p. M-1; *Petroleum Economist* 51, no. 5 (May 1984): 190.

21. *Platt's Oilgram News,* February 23, 1984, pp. 2–3.

22. *Latin American Weekly Report,* June 7, 1985, p. 4.

23. *Oil Daily,* March 19, 1985, p. 6.

24. *Newsday,* September 2, 1985, p. 8.

25. *Latin American Weekly Report,* June 7, 1985, p. 8.

26. *Newsday,* September 2, 1985, p. 8.

27. *El Universal,* May 18, 1985, p. 4.

28. Interview with Manuel Cordero Cuadra, minister-counselor, Nicaraguan embassy, Washington, D.C., May 30, 1985; and *Washington Post,* August 16, 1984.

29. *Facts on File,* February 1, 1985, p. 59.

30. *Platt's Oilgram News,* January 30, 1985, p. 4.

31. Ibid., March 14, 1985, p. 2.

32. *Platt's Oilgram News,* December 3, 1984, p. 4.

33. *Daily Report (Latin America),* March 11, 1985, p. 12.

34. *New York Times,* May 12, 1985, p. 6.

35. *Newsday,* September 2, 1985, p. 8.

36. *New York Times,* April 30, 1985, p. 3.

37. Tass, quoted in ibid.

38. *New York Times,* May 12, 1985, p. 6.

39. *Newsday,* September 2, 1985, p. 4.

40. Ibid., p. 8.

41. *Washington Post,* March 8, 1987, p. A-34.

42. The "Exxon Out of Nicaragua Coalition" announced in April 1987 its plans to picket and leaflet Exxon stations because of its subsidiary refinery operations. The goal of the organization, which claimed that "Exxon fuels communism," was to persuade customers to cut up their credit cards, to stop buying Exxon products, to write protest letters to the corporation's board chairman, and to urge President Reagan to declare that the firm was violating the U.S. embargo against Nicaragua (see *U.S. Oil Week,* April 27, 1987, pp. 1–2).

43. Nicaragua pays Exxon in córdobas for processing Soviet crude. According to spokesmen for the Nicaraguan Foreign Ministry, negotiations over a schedule to convert the córdobas to dollars were continuing in early 1986 (see *Oil Daily,* March 26, 1986, p. 2).

44. Interview with Manuel Cordera Cuadra.

45. *Washington Post,* March 8, 1987, p. A-34.

46. *Latin American Weekly Report,* June 11, 1987, p. 4; *Washington Post,* June 4, 1987, pp. A-1, A-30.

47. *Latin American Weekly Report,* June 11, 1987, p. 4.

48. Ibid.

49. *Financial Times,* June 15, 1987, p. 3.

50. *Washington Post,* August 16, 1984, p. A-29.

8: The Emergence of a Contentious Partner?

1. Martin Needler introduced the pendulum theory—namely, that the presidency has shifted from the left to the center to the right of the political spectrum. See *Politics and Society in Mexico* (Albuquerque, N.M.: University of New Mexico Press, 1971).

2. *Latin American Weekly Report,* May 14, 1987, p. 4.

3. *Economist,* May 23, 1987, p. 78.

4. Ibid., pp. 12–13.

5. Ibid., August 23, 1986, p. 68.

6. *Wall Street Journal,* August 18, 1986, p. 25.

7. Abraham L. Lowenthal, *Partners in Conflict: The United States and Latin America* (Baltimore: Johns Hopkins University Press, 1987), p. 75.

8. *New York Times* poll, "Mexico Survey: October 28–November 4, 1986," pp. 6, 13.

9. Lowenthal, *Partners in Conflict,* p. 67.

10. De Sánchez, Andrés et al., *La renovación política y el sistema electoral mexicano* (Mexico City: Editorial Porrúa, 1987), pp. 110–14.

11. *Latin American Weekly Report,* January 15, 1987, p. 2.

12. Secretaria Técnica, La Comisión Federal Electoral, *La nueva legislación electoral mexicana* (Mexico City: Talleres Gráficos de la Nación, 1987).

13. For an insight into how the SRTPRM reacted negatively to the PRI's opening up the candidate selection process for mayor of Ebano, see the *Wall Street Journal,* June 9, 1987, pp. 1, 26.

14. In a mid-1985 survey on whether the government would respect the results of state and local

elections, 55 percent of the respondents said no, while only 13 percent said yes; see *Excelsior,* July 1, 1985. In a late 1986 poll, 88 percent of respondents in a national sample characterized the condition of the economy as "very bad" or "bad," while 84 percent stated a belief that Mexico's economic conditions would "never" get better (54 percent) or would not improve for at least ten years (30 percent); see the *New York Times* poll, "Mexico Survey," p. 2.

15. Comments at a round-table discussion at Harvard University's Center for International Studies, September 12, 1986.

16. *Latin American Weekly Report,* September 11, 1986, p. 8.

17. Centro de Estudios Económicos del Sector Privado A.C., "La economía subterránea en méxico," *Actividad Económica* 103 (September 1986).

18. *Latin American Weekly Report,* April 9, 1987, p. 8.

19. Jorge G. Castañeda, "Mexico's Coming Challenges," *Foreign Policy* 64 (Fall 1986): 134.

20. *Daily Report (Latin America),* May 6, 1987, p. M-5.

21. Lowenthal, *Partners in Conflict,* p. 76.

22. Dan Nimmo, "Elections as Ritual Drama," *Society* 22, no. 4 (May/June 1985): 31–38.

23. *Wall Street Journal,* May 29, 1987, p. 29.

24. Castañeda, "Don't Corner Mexico!" p. 88.

25. Carlos Rico F., "The Future of Mexico: U.S. Relations and the Limits of the Rhetoric of 'Interdependence' " in *Mexican-U.S. Relations: Conflict and Convergence,* ed. Carlos Vásquez and Manuel García y Griego (Los Angeles, Calif.: UCLA Chicano Studies Research Center Publications, 1983), pp. 127–74.

Bibliography

Books and Monographs

Archivo Histórico-Diplomático. 1st ser.: vols. 33, 36; 2nd ser.: vols. 4, 7.

Atkins, G. Pope. *Latin America in the International Political System*. New York: Free Press, 1977.

Bailey, Norman A., and Cohen, Richard. *The Mexican Time Bomb*. New York: Priority Press, 1987.

Barker, Eugene C. *Mexico and Texas, 1821–1835*. New York: Russell and Russell, 1965.

Brand, D. Donald. *Mexico: Land of Sunshine and Shadow*. Princeton, N.J.: Van Nostrand Co., 1966.

Brandenburg, Frank. *The Making of Modern Mexico*. Englewood Cliffs, N.J.: Prentice-Hall, 1964.

Briseño, Aguilar Jaime. *La lucha de un líder*. Mexico: Los Talleres de Editorial Color, 1980.

Camín Aguilar, Héctor. *Morir en el golfo*. Mexico City: Ediciones Oceano, 1987.

Camp, Roderic A., ed. *Mexico's Political Stability: The Next Five Years*. Boulder, Colo.: Westview Press, 1986.

Carter, Jimmy. *Keeping Faith: Memoirs of a President*. New York: Random House, 1982.

Child, Jack, ed. *Conflict in Central America*. New York: St. Martin's Press, 1986.

Cline, Howard F. *The United States and Mexico*. Rev. ed. New York: Atheneum, 1965.

Delpar, Helen, ed. *Encyclopedia of Latin America*. New York: McGraw-Hill, 1974.

de la Peña, Ricardo, and César Zaqueta. *La estructura del congreso del trabajo, estado, trabajo y capital en méxico: un acercamiento al tema*. Mexico City: Fondo de Cultura Económica. 1984.

de Sánchez, Andrea, et al. *La renovación política y el sistema electoral mexicano*. Mexico City: Editorial Porrúa, 1987.

García y Griego, Manuel, and Carlos Vásquez, eds. *Mexico-U.S. Relations: Conflict and Convergence*. Los Angeles: UCLA Chicano Studies Research Center Publications, 1983.

Grayson, George W. *The Politics of Mexican Oil*. Pittsburgh, Pa.: University of Pittsburgh Press, 1980.

————. *The United States and Mexico: Patterns of Influence*. New York: Praeger, 1984.

Hannaford, Peter. *The Reagans: A Political Portrait*. New York: Coward-McCann, 1983.

Heredia, Rafael Ramírez. *La otra cara del petróleo*. Mexico City: Editorial Diana, 1979.

Holsti, K. J. *International Politics: A Framework for Analysis*. Englewood Cliffs, N.J.: Prentice-Hall, 1983.

Keohane, Robert O., and Joseph S. Nye. *Power and Interdependence: World Politics in Transition*. Boston and Toronto: Little, Brown, 1977.

Kraft, Joseph. *The Mexican Rescue*. New York: Group of Thirty, 1984.

Krasner, Stephen, ed. *International Regimes*. Ithaca, N.Y.: Cornell University Press, 1983.

Levy, Daniel, and Székely, Gabriel. *Mexico: Paradoxes of Stability and Change*. Boulder, Colo.: Westview Press, 1983.

López Moreno, Javier. *La reforma política en méxico*. Mexico City: Centro de Documentación Política, A.C., 1979.

Lowenthal, Abraham L. *Partners in Conflict: The United States and Latin America*. Baltimore: John Hopkins University Press, 1987.

Mancke, Richard B. *Mexican Oil and Natural Gas: Political, Strategic, and Economic Implications*. New York: Praeger, 1979.

Manning, William R. *Early Diplomatic Relations between the United States and Mexico*. New York: Greenwood Press, 1968.

Meyer, Lorenzo, and Josefina Vázquez Zoraida. *México frente a estados unidos: un ensayo histórico, 1776–1980*. Mexico City: El Colegio de México, 1982.

Meyer, Michael C., and Sherman, William L. *The Course of Mexican History*. New York: Oxford University Press, 1983.

Needler, Martin. *Politics and Society in Mexico*. Albuquerque: University of New Mexico Press, 1971.

Newell, Roberto G., and Luis, Rubio F. *Mexico's Dilemma: The Political Origins of Economic Crisis*. Boulder, Colo.: Westview Press, 1984.

Pellicer de Brody, Olga. *México y la revolución cubana*. Mexico City: El Colegio de México, 1972.

Riding, Alan. *Distant Neighbors: A Portrait of the Mexicans*. New York: Alfred A. Knopf, 1985.

Rojas, Francisco. *Hermanemos la idea a la acción*. Mexico City: Mexican Petroleum Institute, 1987.

Rudolph, James D., ed. *Mexico: A Country Study*. American University Foreign Area Studies. Washington, D.C.: Government Printing Office, 1985.

Schmitt, Karl. M. *Mexico and the United States, 1821–1973: Conflict and Coexistence*. New York: Wiley, 1974.

Servan-Schreiber, Jean-Jacques. *Le défi mondial*. Paris: Fayard. 1980.

Smith, Peter H. *Mexico: The Quest for a U.S. Policy*. New York: Foreign Policy Assn., n.d.

Wahlke, John C., et al. *The Legislative System: Explorations in Legislative Behavior*. New York: John Wiley, 1962.

Williams, Edward J. *The Rebirth of the Mexican Petroleum Industry*. Lexington, Mass.: Lexington Books, 1979.

Government Documents

de la Madrid H., Miguel. *Primer informe de gobierno: sector política exterior*. Mexico City: Presidencia de la República, 1983.

Institutional Revolutionary Party. Secretariat of Information and Propaganda. *Who Is Miguel de la Madrid[?]* Mexico City: PRI, 1982.

International Monetary Fund. *Direction of Trade Statistics Yearbook*. Washington, D.C.: IMF, 1970–1986.

López Portillo, José. *México en el ambito internacional*. Vols. 1–4. Mexico City: Secretaria de Promación y Presupuesto, 1981.

Partido Revolucionario Institutional. *Diálogo norte sur*. Mexico City: Tallares Gráficos de la Nación, 1981.

Petróleos Mexicanos. *Anuario estadístico*. Mexico City: Pemex, 1983–1984.

———. *Pemex: Information Bulletin*. June 1984–November 1987.

———. *A Report by Petróleos Mexicanos to Mr. Miguel de la Madrid, the President of Mexico,* about the Industry's Progress and Prospects during the First Two Years of His Administration. Mexico City: Pemex, n.d.

———. *Report from the General Director of Petróleos Mexicanos Mario Ramón Beteta.* Mexico City: Mexican Petroleum Institute, 1975–1986.

———. *Third Evaluation Meeting.* Mexico City: Mexican Petroleum Institute, 1987.

Poder Ejecutivo Federal. *Programa nacional de energéticos, 1984*–1988. Mexico City: SEMIP, 1984.

Presidencia de la República Mexicana. *Informe de gobierno que rinde ante el congreso de la unión.* Mexico City, 1976–1986.

———. *Plan mundial de energía: posición ante las naciones unidas.* Mexico City: Coordinación General de Comunicación Social, 1979.

Republic of Mexico. *Cámara de Diputados. Directorio: 1982–1985.* Mexico City: Congreso de la Unión, n.d.

Republic of Venezuela. Ministerio de Información y Turismo. *La cooperación internacional de venezuela: solidaridad en acción.* Caracas: Imprenta Nacional, 1982.

———. Ministry of Energy and Mines. *La cooperación energética de venezuela/The Energy Cooperation of Venezuela.* Caracas: Ediciones del Ministerio de Energía y Minas, 1981.

Secretaria Técnica de la Comisión Federal Electoral. *La nueva legislación electoral mexicana.* Mexico City: Talleres Gráficos de la Nación, 1987.

Trinidad and Tobago's Economic Assistance to the Caricom Member Countries: A Brief Review—1979–1982. Pamphlet provided by the Embassy of Trinidad and Tobago, Washington, D.C., n.d.

U.S. Department of Commerce. Bureau of Economic Analysis. *Survey of Current Business.* Washington, D.C.: Government Printing Office, 1970–1987.

———. Office of Research and Analysis, U.S. Travel Service. *Foreign Visitor Arrivals, 1966–1976.* Washington, D.C.: Government Printing Office.

———. Social and Economic Statistics Administration, Bureau of the Census. *Statistical Abstract of Latin America.* Washington, D.C.: Government Printing Office, 1986.

U.S. Department of Energy. DOE Contract No. DE-AC01-82EP31403, signed August 23, 1982.

U.S. Department of Justice. Criminal Division. *The Foreign Agents Registration Act of 1938 as Amended and the Rules and Regulations Prescribed by the Attorney General.* Washington, D.C.: Government Printing Office, 1986.

———. Immigration and Naturalization Service. *1984 Statistical Yearbook of the Immigration and Naturalization Service.* Washington, D.C.: Government Printing Office.

U.S. Department of State. "The Mexican-Venezuelan Oil Facility—An Assessment." Washington, D.C., May 13, 1983. Mimeo.

———. Bureau of Public Affairs. *Background on the Caribbean Basin Initiative.* Special Report No. 97. Washington, D.C: Government Printing Office, 1982.

———. Bureau of Public Affairs. *Caribbean Basis Initiative.* Current Policy No. 370. Washington, D.C.: Government Printing Office, 1982.

U.S. Embassy, Mexico City. "Economic Trends Report." February 1987. Mimeo.

U.S. Federal Communications Commission. *Statistics of Communications: Common Carriers.* Washington, D.C.: Government Printing Office, 1971–1983.

U.S. Foreign Broadcast Information Service. *Daily Report (Latin America),* 1978–1987.

Articles and Chapters in Books

Baer, M. Delal. "Mexico: Ambivalent Ally," *Washington Quarterly* 10 (Summer 1987): 103–13.

Bagley, Bruce Michael. "Mexican Foreign Policy in the 1980's: A New Regional Power." *Current History.* 80 (November 1981): 353–56.

————. "Contadora: The Failure of Diplomacy." *Journal of Inter-American Studies and World Affairs* 28 (Fall 1986): 1–32.

Blanksten, George I. "Foreign Policy of Mexico." In *Foreign Policy in World Politics*, ed. Roy C. Macridis. 2d ed.: Englewood Cliffs, N.J.: Prentice-Hall, 1963, pp. 311–33.

Bond, Robert D. "Venezuelan Policy in the Caribbean Basin." In *Central American: International Dimensions of the Crisis*, ed. Richard E. Feinberg. New York; Holmes & Meier, 1982.

"Caribbean Oil Deal." *South* 2 (November 1980).

Castañeda, Jorge. "Revolution and Foreign Policy: Mexico's Experience." *Political Science Quarterly* 78 (September 1963): 391–417.

Castañeda, Jorge G. "Don't Corner Mexico!" *Foreign Policy* 60 (Fall 1985): 75–90.

————. "Mexico at the Brink." *Foreign Affairs* 64 (Winter 1985–86): 287–303.

————. "Mexico's Coming Challenges." *Foreign Policy* 64 (Fall 1986): 120–39.

Centro de Estudios Económicos del Sector Privado, A.C., "La economía subterránea en méxico," *Actividad Económica* 103 (September 1986): 1–11.

Cornelius, Wayne A. "The Political Economy of Mexico under de la Madrid: Austerity, Routinized Crisis, and Nascent Recovery." *Mexican Studies/Estudios Mexicanos* 1, no. 1 (Winter 1985): 83–124.

Cuevas Cancino, Franciso. "The Foreign Policy of Mexico." In *Foreign Policies in a World of Change*, ed. Joseph E. Black and Kenneth W. Thompson. New York: Harper & Row, 1963.

Deagle, Edwin A., Jr. "The United States National Security Policy and Mexico." In *U.S.-Mexican Relations: Economic and Social Aspects*, ed. Clark W. Reynolds and Carlos Tello. Stanford, Calif.: Stanford University Press, 1983.

Grayson George W. "The Maple Leaf, The Cactus, and the Eagle: Energy Trilateralism. *Inter-American Economic Affairs* 34, no. 4 (Spring 1981): 49–75.

————. "Mexico: The Oil Glut and Structural Reform." *Washington Quarterly* 9, no. 3 (Summer 1986): 153–64.

————. "Middle-Class Agitation in Quaking Mexico." *Washington Post*, December 8, 1985, pp. B-1, B-4.

————. "Nicaragua: Soviets Aid with Oil Supplies." *Petroleum Economist* 52, no. 7 (July 1985): 251–53.

————. "The San José Oil Facility: South-South Cooperation." *Third World Quarterly* 7, no. 2 (April 1985): 390–409.

————. "An Overdose of Corruption: The Domestic Politics of Mexican Oil." *Caribbean Review* 13, no. 3 (Summer 1984): 22–24, 46–49.

————. "The U.S.-Mexican Gas Deal and What We Can Learn from It." *Orbis* 78, no. 454 (February 1980): 53–56.

————. "Venezuela and the Puerto Ordaz Agreement." *Inter-American Economic Affairs* 38, no. 3 (Winter 1984): 49–73.

Halperin, Maurice. "Mexico Shifts Her Foreign Policy," *Foreign Affairs* 10 (June 1941): 207–21.

Holsti, K. J. "National Role Conceptions in the Study of Foreign Policy." *International Studies Quarterly* 14 (1970): 233–309.

Holt, Donald D. "Why the Bankers Suddenly Love Mexico." *Fortune*, July 16, 1979, pp. 138–45.

"The Inaugural Address of John Fitzgerald Kennedy." *The Kennedy Presidential Press Conferences*. New York: Earl M. Coleman, 1978.

Keohane, Robert O. "Theory of World Politics: Structural Realism and Beyond." In *Neorealism and Its Critics*, ed, Keohane. New York: Columbia University Press, 1986.

Klein, Dianne. "Repainting the Picture: Mexico turns to PR to Improve Image in U.S." *Houston Chronicle,* April 20, 1986, p. 30.

Koehler, Wallace C., Jr., and Segal, Aaron L. "Prospects for North American Energy Cooperation." *USA Today,* May 1980, pp. 40–43.

Lake, David. "International Economic Structures and American Foreign Economic Structures, 1887–1934." *World Politics* (July 1983): 517–43.

Latell, Brian. "Mexico at the Crossroads: The Many Crises of the Political System." Stanford, Calif.: Hoover Institution of Stanford University, June 16, 1986.

"Mexican Oil Boom Still Held Back by Debt Problems." *Latin American Economic Report,* November 4, 1977, p. 197.

"Mexico's New Muscle." *Newsweek,* October 1, 1979, pp. 26–32.

Meyer, Herbert E. "Why a North American Common Market Won't Work—Yet." *Fortune,* September 10, 1979, pp. 118–24.

Nimmo, Dan. "Elections as Ritual Drama." *Society* 22, no. 4 (May/June 1985): 31–38.

Ojeda, Mario. "The Future of Mexico-U.S. Relations." In *U.S.-Mexico Relations: Economic and Social Aspects,* ed. Clark W. Reynolds and Carlos Tello. Stanford, Calif.: Stanford University Press, 1983.

O'Leary, Michael. "Linkage between Domestic and International Politics in Undeveloped Nations." In *Linkage Politics,* ed. James N. Rosenau. New York: Free Press, 1969.

Peaslee, Amos J. "Political Constitution of the United States of America and Mexico." In *Constitutions of Nations,* vol. 2. The Hague: Martinus Nijhoff, 1956.

Poitras, Guy. "Mexico's 'New' Foreign Policy," *Inter-American Economic Affairs* 28 (Winter 1974): 59–74.

Purcell, Susan Kaufman. "Demystifying Contadora." *Foreign Affairs* 64 (Fall 1985): 74–95.

———. "Mexico-U.S. Relations: Big Initiatives Can Cause Big Problems." *Foreign Affairs* 60, no. 2 (Winter 1981–1982): 379–92.

Rico, Carlos F. "The Future of Mexican—U.S. Relations and the Limits of the Rhetoric of 'Interdependence.' " In *Mexican-U.S. Relations: Conflict and Convergence,* ed. Carlos Vásquez and Manuel García y Griego. Los Angeles: UCLA Chicano Studies Research Center Publications, 1983, pp. 127–74.

Riding, Alan. "The Mixed Blessings of Mexican Oil." *New York Times Magazine,* January 11, 1981, pp. 22–25.

Samaniego, Fidel. "Cinco, los miembros del gabinete con más posibilidades de ser candidatos del pri," *El Universal,* March 25, 1987, pp. 10–11.

Shapira, Yoram. "Mexico's Foreign Policy under Echeverría: A Retrospect." *Inter-American Eocnomic Affairs* 31 (Spring 1978): 29–61.

Smith, Peter H. "Uneasy Neighbors: Mexico and the United States." *Current History* 86 (March 1987): 97–134.

Székely, Gabriel. "México y el petróleo: 1981–1985." *La Jornada,* September 1, 1985.

Sweeney, Jack. "Finance to Flow from Accord . . . Against Promises." *Business Mexico* 4 (March 1987): 24–29.

"Trinidad's Oil Facility Gets Underway." *Caribbean and West Indies Chronicle* 96, no. 1557 (1980): 15.

"Una labor social sin presendentes del trabajador petrolero." *Siempre!* June 20, 1984, pp. 43–46.

Weil, Dan. "Mexico, Venezuela Seen Likely to Renew San José Accord Although Its Significance has Lessened." AP–Dow Jones Wire Service, July 17, 1986, p. 2.

Wellhausen, Edward J. "The Agriculture of Mexico." *Scientific American* 235 (September 1976): 128–30ff.

Williams, Edward J. "Mexico, Oil and OPEC." *Latin American Digest* 11–12 (Fall–Winter 1977–1978): 4–6.
Wonder, Edward R., and Elliott, J. Mark. "Caribbean Energy Issues and U.S. Policy." In *Western Interests and U.S. Options in the Caribbean Basin,* ed. James R. Greene and Brent Scowcroft. Boston: Oelgeschlager, Gunn & Hain, 1984.

Magazines and Newspapers

Business Week. February 1983–February 1985.
Comercio exterior de méxico.
Congressional Quarterly Almanac. 96th Cong., 2d sess. Vol. 36. Washington, D.C.: Congressional Quarterly, Inc., 1980.
Economist (London). 1983–1987.
Excelsior. 1977–1987.
Facts on File. 1979–1987.
Financial Times (London). 1983–1987.
Fortune. 1979–1987.
International Studies Quarterly 14, no. 3 (1970): 233–309.
Journal of Inter-American Studies and World Affairs. 1985–1987.
Keesing's Contemporary Archives. 1979–1987.
Latin American Weekly Report. 1981–1987.
Latin American Political Report. (25 May 1979): 154.
Latin America Regional Report: Mexico and Central America. 1980–1987.
Middle East Economic Survey. July 25, 1975–December 31, 1984.
Newsday. September 2, 1985.
New York Times. 1970–1987.
Oil and Gas Journal. 1970–1987.
Oil Daily. 1975–1987.
OPEC Bulletin. 1980–1987.
Petroleum Economist. 1970–1987.
Petroleum Intelligence Weekly. 1970–1987.
Platt's Oilgram News. 1979–1987.
Platt's Oilgram Price Report. May 18, 1987.
Proceso. 1976–1987.
El Tiempo. February 21, 1977.
El Universal. May 18, 1985 and June 2, 1986.
U.S. Oil Week, April 27, 1987, pp. 1–2.
Wall Street Journal. 1970–1987.
Washington Post. 1970–1987.
World Oil. August 15, 1979–August 15, 1982.

Unpublished Works

Bagley, Bruce Michael. "Regional Powers in the Caribbean Basin: Mexico, Venezuela, and Colombia." Occasional Paper No. 2, School of Advanced International Studies, Johns Hopkins University, January 1983.
Bailey, John J. "What Impact Will Major Groups Have on Policy-Making?: Trends in Government/Business Relations in Mexico." Issue Paper 5 Prepared for Working Group on Mexico: The Next Five Years.
Border Gas. Press release. Houston, Texas, October 24, 1984.

Council of the Mexico-U.S. Business Committee. "U.S. Council of the Mexico-U.S. Business Committee, 1984–85 Report." New York, n.d.

Eschbach, Cheryl. Ph.D. Candidate in Government, Harvard University. Undated memorandum.

Fagen, Richard R., and Nau, Henry R. "Mexican Gas: The Northern Connection." Presented at the Conference on the United States, U.S. Foreign Policy, and Latin American and Caribbean Regimes, Washington, D.C., March 27–31, 1978.

Gutiérrez Kirchner, Alfredo. "Presencia, funciones y actividades de petróleos mexicanos en los estados unidos de america," 1983.

The Hannaford Company, Inc. "Camerena Slaying Suspects Seized as Mexico's Drug Crackdown Continues." News release. Washington, D.C., February 6, 1986.

———. "Confessions Implicates Camerena Slaying Suspects." News release. March 27, 1986.

———. "Mexico Police Net Record $80 Million Drug Haul." News release. April 9, 1986.

International Petroleum Encyclopedia. 1965–1976. Tulsa, Okla: Penwell Publishing Co.

Menges, Constantine C. "Concurrent Mexican Foreign Policy Revolution in Central America and the United States." Prepared for the Hudson Institute Conference, Washington, D.C., June 1980.

New York Times Poll. "Mexico Survey: October 28–November 1986."

Partido Revolucionario Institutional, Instituto de Estudios Políticos, Económicas y Sociales. Comisión de Energéticos. "Energía y sector externo." Draft for discussion, September 22, 1982.

Ronfeldt, David, and Sereseres, Caesar. "Immigration Issues Affecting the U.S.-Mexican Relations." Presented to the Brookings Institution–El Colegio de México Symposium on Structural Factors Contributing to Current Patterns of Migration in Mexico and the Caribbean Basin, Washington, D.C., June 28–30, 1978.

U.S. Council of the Mexico-U.S. Business Committee. "Communiqué." Presented at 41st Plenary Meeting, Quintana Roo, Mexico, November 8, 1986. Draft communiqué, January 21, 1981.

Interviews

Arana, Deputy Lic. Fernando Ortiz. Secretary of Electoral Action, PRI. Mexico City, May 20, 1987.

Arreaza, René. Petróleos de Venezuela, S.A. Caracas, August 7, 1984.

Cameron, Jared. Vice-President, The Hannaford Company, Washington, D.C., June 5, 1987.

Cuadra, Manuel Cordero. Minister-Counselor, Nicaraguan embassy. Washington, D.C., May 30, 1985.

de la Madrid Hurtado, Miguel. President of Mexico. Mexico City, November 11, 1987.

Fogt, Brent. Mexico Division, U.S. Department of Commerce. Washington, D.C., June 3, 197.

Fritts, Robert E. Diplomat-in-Residence, College of William and Mary. Williamsburg, Va., September 8, 1986.

Gil, Julio. Minister-Counselor, Embassy of Venezuela, Washington, D.C., June 5, 1984.

Hernández Valentín. Venezuelan Ambassador to the United States and Minister of Energy and Mines (1974–1979). Washington, D.C., May 30, 1984.

Moran, Mark E. General Counsel, The Hannaford Company. Washington, D.C., May 22, 1986 and July 24, 1986; Mexico City, November 11, 1987.

Pérezgasga Tovar, Flavio. Subdirector for Planning and Coordination Petróleos Mexicanos. Mexico City, August 15, 1985.

Stanley, Kenneth. Industry Analysis Division. Federal Communications Commission. Washington, D.C., June 3, 1987.

Woodrow, Karen. U.S. Department of Commerce, Bureau of the Census. Washington, D.C., June 2, 1987.

Index

Pitt Latin American Series

Cole Blasier, Editor

The Overthrow of Allende and the Politics of Chile, 1964–1976
Paul E. Sigmund

Panajachel: A Guatemalan Town in Thirty-Year Perspective
Robert E. Hinshaw

Peru and the International Monetary Fund
Thomas Scheetz

Primary Medical Care in Chile: Accessibility Under Military Rule
Joseph L. Scarpaci

Rebirth of the Paragayan Republic: The First Colorado Era, 1878–1904
Harris G. Warren

Social Security

The Politics of Social Security in Brazil
James M. Malloy

Social Security in Latin America: Pressure Groups, Stratification, and Inequality
Carmelo Mesa-Lago

Other Studies

Adventurers and Proletarians: The Story of Migrants in Latin America
Magnus Mörner, with the collaboration of Harold Sims

Authoritarianism and Corporatism in Latin America
James M. Malloy, Editor

Authoritarians and Democrats: Regime Transition in Latin America
James M. Malloy and Mitchell. A Seligson, Editors

Female and Male in Latin America: Essays
Ann Pescatello, Editor

Latin American Debt and the Adjustment Crisis
Rosemary Thorp and Laurence Whitehead, Editors

Public Policy in Latin America: A Comparative Survey
John W. Sloan

Selected Latin American One-Act Plays
Francesca Colecchia and Julio Matas, Editors and Translators

The State and Capital Accumulation in Latin America: Brazil, Chile, Mexico
Christian Anglade and Carlos Fortin, Editors

Transnational Corporations and the Latin American Automobile Industry
Rhys Jenkins